Springer Undergraduate Mathematics Series

W0080828

The Springer Undergraduate Mathematics Series (SUMS) is a series designed for undergraduates in mathematics and the sciences worldwide. From core foundational material to final year topics, SUMS books take a fresh and modern approach. Textual explanations are supported by a wealth of examples, problems and fully-worked solutions, with particular attention paid to universal areas of difficulty. These practical and concise texts are designed for a one- or two-semester course but the self-study approach makes them ideal for independent use.

David Hernandez • Yves Laszlo

Introduction to Galois Theory

 Springer

David Hernandez
Institut de Mathématiques de Jussieu-Paris
Rive Gauche
Université Paris Cité
Paris, France

Yves Laszlo
Institut de mathématiques d'Orsay
Université Paris-Saclay
Orsay, France

ISSN 1615-2085   ISSN 2197-4144   (electronic)
Springer Undergraduate Mathematics Series
ISBN 978-3-031-66181-5   ISBN 978-3-031-66182-2   (eBook)
https://doi.org/10.1007/978-3-031-66182-2

The original submitted manuscript has been translated into English. The translation was done using artificial intelligence. A subsequent revision was performed by the author(s) to further refine the work and to ensure that the translation is appropriate concerning content and scientific correctness. It may, however, read stylistically different from a conventional translation.

Translation from the French language edition: "Introduction à la théorie de Galois" David Hernandez and Yves Laszlo, © 2012 Les Éditions de l'École Polytechnique. Published by Les Éditions de l'École polytechnique. All rights reserved.

This Springer imprint is published by the registered company Springer Nature Switzerland AG
The registered company address is: Gewerbestrasse 11, 6330 Cham, Switzerland

If disposing of this product, please recycle the paper.

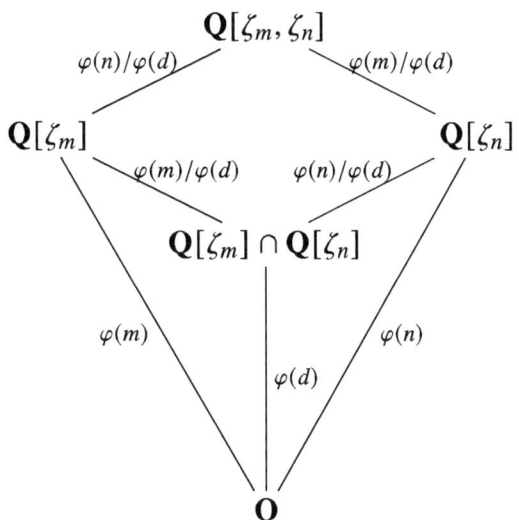

Preface

The purpose of this book is to show how closely related two seemingly unrelated fields—group theory and field extensions—are. This deep connection, highlighted in the nineteenth century by Galois,[1] allows for significant results in arithmetic.[2] Even though, due to time constraints, we could hardly present modern results, Galois theory and its extensions currently hold a central place in Mathematics. A simple search for the word "Galois" in the MathSciNet mathematical database returns 25,594 references and the number of occurrences up to the last stabilized year of the database is an essentially strictly increasing function of time with 719 references in 2022! However, our understanding of Galois groups, especially of number fields, remains very partial, even though spectacular progress has been made in the last 50 years.

This book is derived from a course given at the École Polytechnique (Palaiseau, France), first by the second author and then by the first. In this course, we preferred to break with tradition by only outlining the solutions to classic and age-old problems provided by Galois theory (constructibility with a ruler and compass, for example), to go further in the presentation of powerful algebraic methods (introduction to the reduction modulo p of Galois groups (Chap. 10)) or recent results (some results from inverse Galois theory (Sect. 11.6)). We also did not seek to develop sophisticated methods for the algorithmic calculation of Galois groups, which exist, but which, in our opinion, are rather "expert" problems. We also did not cover the theory of resolvents. We have scattered exercises throughout the text, which are mostly very simple but will allow you to "get your hands dirty" and check if the concepts are assimilated. We encourage the reader to only consult the proof hints in Chap. 13 as a last resort. Review exercises, mostly from exam topics given at the École Polytechnique, are gathered in Chap. 12 and solved in Chap. 13.

---

[1] A recent historical presentation can be found in [Ehr11].

[2] The "queen of sciences" as Gauss said.

Fig. 1 Alexandre
Grothendieck (1928–2014).
Author: Paul R. Halmos.
Source: Paul R. Halmos
photograph collection, The
Dolph Briscoe Center for
American History, The
University of Texas at Austin

Regarding the bibliography, one can refer to the best-seller by I. Stewart [Ste15], and the beautiful books by A. Chambert-Loir [Cha05] and R. Elkik [Elk02]. To go further, especially in the study of separability, [Bou23] is a classic.

There are no prerequisites beyond basic (mostly linear) algebra. In particular, we have avoided the tensor product, which alone would have made many proofs significantly more natural. Alas, time was lacking.[3]

Galois theory has been widely generalized and is in fact only a particular case of Alexander Grothendieck's (Fig. 1) vast theory of faithfully flat descent, revealed in [Gro71], which, in a sense, is simpler and more geometric.

Although Grothendieck's work is hardly accessible to the audience of this book, his geometric point of view is exposed in the beautiful book by R. and A. Douady [Dou20], a work whose reading cannot be recommended too highly. It explains the analogy between number fields and Riemann surfaces and the Galois correspondence between field extensions and finite ramified coverings of these surfaces! It also addresses the rich and active theory of Grothendieck's *children's drawings*, which bridges the theory of Riemann surfaces and arithmetic via the study of the Galois group of $\overline{\mathbf{Q}}$ over $\mathbf{Q}$.

This course is therefore an invitation to travel rather than an exhaustive exposition, which would have required much more space.

From a technical point of view, we have especially limited ourselves to perfect fields, which has allowed us to avoid discussions of separable extensions. It seemed to us not to hinder the understanding of the methods, especially since this framework covers many current problems. We have not restricted ourselves to fields of characteristic zero in order to have a theory that encompasses the case of finite fields, which, as we will see in Chap. 10, is very useful for calculating the Galois groups of fields of characteristic zero.

---

[3] The course on which this book is based was given for second-year students at the École Polytechnique. The constraints were to give a presentation of Galois theory in 14 hours of lectures without any prerequisites in algebra, hence the lack of time mentioned above. One of the aims of this book is to show that one can indeed quickly get to the essentials.

We have deliberately sought to go as quickly as possible in the proofs, as long as they remained "natural," without seeking to unnecessarily generalize them (cf. for example the discussions on algebraic integers).

May the beauty and power of this wonderful theory touch the reader.

Paris, France   David Hernandez
Orsay, France   Yves Laszlo
May 2024

Prologue

On the night of May 29, 1832, Évariste Galois (Fig. 2) knew his death was near. He wrote a testamentary letter[4] addressed to his friend Auguste Chevalier, a facsimile of which is presented in Fig. 3.

The first leaf of the aforementioned letter begins as follows. Even if the style seems a bit obscure, the reader will first recognize the definition of a normal subgroup (Definition 2.2.1) then the theorem of resolvability of algebraic equations (Theorem 9.4.2) and the fact that $5 \cdot 4 \cdot 3$ is the smallest cardinality of a simple non-abelian group (namely A_5).

My Dear Friend,

I have made several new analyses. Some concern the theory of Equations, others Integral functions. In the theory of equations, I have investigated in which cases the equations were solvable by radicals: this gave me the opportunity to delve deeper into this theory, and to describe all the possible transformations on an equation even when it is not solvable by radicals.

Fig. 2 Évariste Galois (1811–1832). Author unknown. Source: Wikimedia Commons

---

[4] See [Gal62] or https://www.ias.ac.in/article/fulltext/reso/004/10/0093-0100 for an English translation.

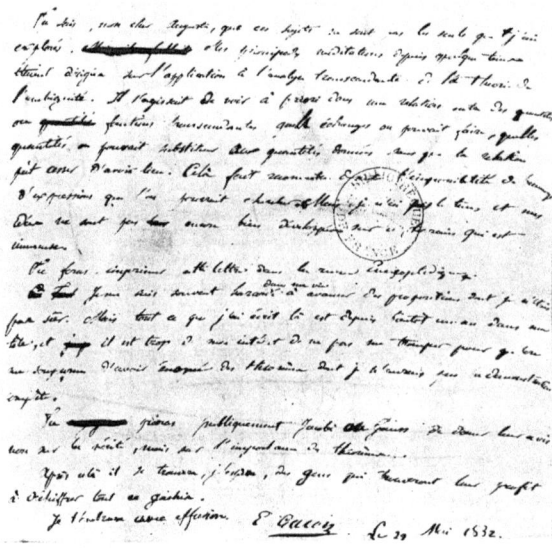

Fig. 3 Facsimilie of a letter of Galois to Auguste Chevalier. Source: Iyanaga, Shokichi, ガロアの
時代 ガロアの数学 第一部 時代篇, ©Springer-Verlag Tokyo, 1999

*All this could make up three papers. The first one is written, and despite what
Poisson said, I maintain it with the corrections I have made. The second contains
quite curious applications of the theory of equations. Here is the summary of the
most important points:*

*According to propositions II and III of the 1st Paper, we see a significant
difference between adding to an equation one of the roots of an auxiliary
equation, or adding them all. In both cases, the group of the equation is divided
by the addition into groups such that one passes from one to the other by the
same substitution. But the condition that these groups have the same substitutions
certainly occurs only in the second case. This is called proper decomposition. In
other words, when a group G contains another group H, the group G can be
divided into groups that are each obtained by operating on the permutations of
H by the same substitution, so $G = H + HS + HS' + \ldots$ and it can also decompose
into groups that all have the same substitutions so that $G = H + TH + T'H + \ldots$*
*These two types of decomposition do not usually coincide. When they do, the
decomposition is said to be proper. It is easy to see that when the group of
an equation is not susceptible to any proper decomposition, no matter how the
equation is transformed, the groups of the transformed equations will always
have the same number of permutations.*

*On the contrary, when the group of an equation is susceptible to a proper
decomposition so that it is divided into m groups of n permutations, the given
equation can be solved by means of two equations: one will have a group of m
permutations, the other one of n permutations.*

Therefore, when all the possible proper decompositions on the group of an equation have been exhausted, we arrive at groups that can be transformed, but whose permutations will always be in the same number.
If these groups each have a prime number of permutations, the equation will be solvable by radicals. Otherwise, not. The smallest number of permutations that an indecomposable group can have when this number is not prime is $5 \cdot 4 \cdot 3$.

And let's end this historic tribute with the moving ending of this last letter, Galois having to die tragically a few hours later following a duel.

[. . .]I have often ventured in my life to advance propositions of which I was not sure. But everything I have written there has been in my head for almost a year, and it is too much in my interest not to be mistaken for anyone to suspect me of having stated theorems of which I would not have the complete demonstration. You will publicly ask Jacobi and Gauss to give their opinion, not on the truth, but on the importance of the theorems. I kiss you with effusion.

Contents

Chapter 1
Invitation to Galois Theory

We will sketch, in a rather informal way, two historically important successes of Galois theory. In this invitation, we will only use the well-known fact that the data of a subfield k of K equips K with the structure of a k-vector space. The dimension, finite or not, is denoted $[K : k]$ and is also called the degree of the *extension* K/k.

The beginning of the proper course starts in Chap. 2.

1.1 Construction With a Straightedge and Compass

We identify the (oriented) Euclidean plane with $\mathbf{C}$, equipped with the usual norm $\|z\|=|z|$.

For two distinct points A, B in $\mathbf{C}$, the unique line passing through A and B is denoted $\langle A, B \rangle$. For A a point in $\mathbf{C}$ and R a positive real number, the circle with center A and radius R is denoted $C(A, R)$. A geometric object that is a line or a circle is called a "circle-line".

Definition 1.1.1 We will say that a point $P \in \mathbf{C}$ is *constructible* if there exists a finite sequence of distinct points $P_0, \ldots, P_N = P$ such that $P_0 \in \{0, 1\}$ and for all $n < N$ the point P_{n+1} is one of the points of a finite intersection of two circle-lines of type $(\langle P_\alpha, P_\beta \rangle), 0 \leq \alpha < \beta \leq n$ or $C(P_\gamma, |P_\alpha - P_\beta|), 0 \leq \alpha < \beta \leq n, \gamma \leq n$.

In other words, we first decide that 0, 1 are constructible. Then, recursively, given a set of constructible points, we construct the lines passing through two distinct constructible points, or a circle centered on one of these points, with a radius equal to the distance between two constructible points: this defines the admissible circle-lines. The constructible points of rank $n + 1$ are the constructible points of rank n together with the finite intersections between two admissible circle-lines.

For example, the complex number $I = \sqrt{-1}$ is constructible.

© The Author(s), under exclusive license to Springer Nature Switzerland AG 2024
D. Hernandez, Y. Laszlo, *Introduction to Galois Theory*, Springer Undergraduate
Mathematics Series, https://doi.org/10.1007/978-3-031-66182-2_1

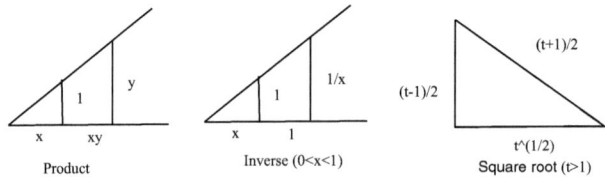

Fig. 1.1 Useful constructions

The reader will remember the theorems of Thales and Pythagoras and will show the properties displayed in Fig. 1.1.

Exercise 1.1.2 The set of constructible reals is a subfield of $\mathbf{R}$ (in particular, it contains the rationals). A positive real number is constructible if and only if its square root is. A complex number z is constructible if and only if its real and imaginary parts are, so that the constructible complex numbers form a subfield of $\mathbf{C}$.

Later (Theorem 8.5.1) we will prove the following result.

Theorem 1.1.3 (Wantzel[a]) *A complex number z is constructible if and only if there exists a finite sequence of fields $L_0 = \mathbf{Q} \subset L_1 \subset \cdots \subset L_n$ such that for each i, $[L_{i+1} : L_i] = 2$ and $z \in L_n$.*

  When this condition is satisfied, we have in particular that $[\mathbf{Q}[z] : \mathbf{Q}]$ is finite and is a power of 2.

---

[a]1814–1848, Lecturer at Polytechnique.

For example, π being transcendental (Sect. 11.3), we conclude the impossibility of squaring the circle: we cannot construct with a straightedge and compass a square of the same area as the unit disk (Fig. 1.2).

We can also deduce for example that we cannot construct with a straightedge and compass a regular heptagon. Indeed, otherwise, the dimension of $\mathbf{Q}[\exp(\frac{2i\pi}{7})]$ over $\mathbf{Q}$ would be a power of 2. However, we will prove the following proposition (Theorem 8.3.8).

Fig. 1.2 Karl Friedrich
Gauss (1777–1855). Author
unknown. Source:
ETH-Bibliothek Zürich,
Bildarchiv, Dia 326-268
(ETH, http://doi.org/10.3932/
ethz-a-001421507)

Fig. 1.3 Leonhard Euler
(1707–1783). Author
unknown. Source:
ETH-Bibliothek Zürich,
Bildarchiv, Dia 326-196

Proposition 1.1.4 (Gauss) *We have* $[\mathbf{Q}[\exp(\frac{2\mathrm{i}\pi}{n})], \mathbf{Q}] = \varphi(n)$, *where φ is Euler's[b] totient function and* $\mathbf{Q}[\exp(\frac{2\mathrm{i}\pi}{n})]$ *is the field generated by* $\exp(\frac{2\mathrm{i}\pi}{n})$, *which is also the set of polynomials with rational coefficients in* $\exp(\frac{2\mathrm{i}\pi}{n})$.

[b]Figure 1.3.

The formula $\varphi(7) = 7 - 1 = 6$ then implies the result.

Generally, therefore, if the regular polygon with n sides is constructible, $\varphi(n)$ is a power of 2. We will see that then n is the product of a power of 2 by a finite number of Fermat primes F_m. We recall here that the Fermat number F_m is $2^{2^m} + 1$.

This result is due to Gauss (Fig. 1.2). These results *do not* involve Galois theory. The converse was conjectured, it seems, by Gauss.

As always, he had guessed right:

Theorem 1.1.5 (Gauss–Wantzel) *The converse is true: if n is a product of a power of 2 and a finite number of distinct Fermat primes, then the regular polygon with n sides is constructible.*

In fact, the proof almost provides an algorithm to construct a regular polygon with n sides (when it is possible!): one must decompose n into prime factors *and* find a generator of the cyclic group $(\mathbf{Z}/p\mathbf{Z})^*$. Note that we have $F_0 = 3, F_1 = 5, F_2 = 17, F_3 = 257, F_4 = 65537$ and they are all prime. While the constructions of equilateral triangles, squares, and regular pentagons are elementary, that of the regular polygon with 17 sides is less obvious.

Let us first recall the construction of the regular pentagon displayed in Fig. 1.4 (known to Ptolemaeus, first century AD, Fig. 1.5), a simple consequence of the elementary formula

$$\cos\left(\frac{2\pi}{5}\right) = \frac{\sqrt{5} - 1}{4}$$

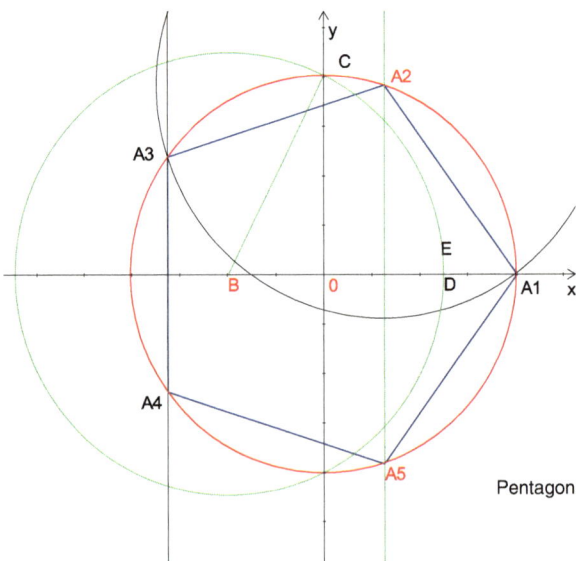

Fig. 1.4 Construction of the regular pentagon

Fig. 1.5 Claudius
Ptolemaeus (c. 90–168).
Author unknown. Source:
ETH-Bibliothek Zürich,
Bildarchiv, Dia 326-579
(ETH, http://doi.org/10.3932/
ethz-a-001421817)

Fig. 1.6 Construction of the
regular heptadecagon
(Wikipedia, https://upload.
wikimedia.org/wikipedia/
commons/3/31/01-
Siebzehneck-Richmond.svg)

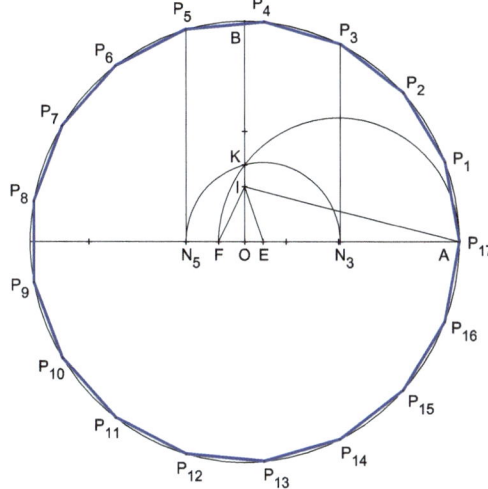

Gauss, again, gave a construction of the 17-sided polygon presented in Fig. 1.6.
Here we already have a rather complicated formula

$$16 \cos\left(\frac{2\pi}{17}\right)$$

$$= -1 + \sqrt{17} + \sqrt{34 - 2\sqrt{17}} + \sqrt{68 + 12\sqrt{17} - 4\sqrt{34 - 2\sqrt{17}} - 8\sqrt{34 + 2\sqrt{17}}},$$

which is derived from Galois theory, and which allows for an actual construction.

On the other hand, F_5 is divisible by 641 (Euler). We do not know if F_{33} is prime,
while we know that $F_{2478782}$ is not: little is known about the primality of Fermat
numbers.

The converse, however, involves Galois theory: it is a consequence of the
calculation of the Galois group $\text{Gal}(\mathbf{Q}[\exp(\frac{2\mathrm{i}\pi}{n})], \mathbf{Q})$ (cf. Theorem 8.3.10).

Fig. 1.7 Gerolamo Cardano
(1501–1576). Self-portrait.
Source: Wikimedia Commons
(Wikipedia, https://upload.
wikimedia.org/wikipedia/
commons/9/97/
Jerôme_Cardan.jpg)

Fig. 1.8 Tartaglia
(1499–1557). Author
unknown. Source: Wikimedia
Commons (Wikipedia,
https://upload.wikimedia.org/
wikipedia/commons/0/0b/
Niccolò_Tartaglia.jpg)

1.2   Solving Equations

Everyone knows the solutions to the quadratic equation $x^2 - a = 0$, $a \in \mathbf{C}$, namely
$x = \pm\sqrt{a}$. In general, for the degree n equation, a clever translation of the variable
cancels the degree $n-1$ term. In degree 3, we are therefore dealing with the equation
$x^3 + ax + b = 0$ whose solutions were bought in the sixteenth century by Cardan
(Fig. 1.7) from the mathematician Tartaglia (Fig. 1.8), but were probably known to
del Ferro.[3] They are written as

$$x_1 = \sqrt[3]{-\frac{b}{2} + \sqrt{\left(\frac{a}{3}\right)^3 + \left(\frac{b}{2}\right)^2}} + \sqrt[3]{-\frac{b}{2} - \sqrt{\left(\frac{a}{3}\right)^3 + \left(\frac{b}{2}\right)^2}}$$

$$x_2 = j\sqrt[3]{-\frac{b}{2} + \sqrt{\left(\frac{a}{3}\right)^3 + \left(\frac{b}{2}\right)^2}} + j^2\sqrt[3]{-\frac{b}{2} - \sqrt{\left(\frac{a}{3}\right)^3 + \left(\frac{b}{2}\right)^2}}$$

$$x_3 = \bar{j}\sqrt[3]{-\frac{b}{2} + \sqrt{\left(\frac{a}{3}\right)^3 + \left(\frac{b}{2}\right)^2}} + \bar{j}^2\sqrt[3]{-\frac{b}{2} - \sqrt{\left(\frac{a}{3}\right)^3 + \left(\frac{b}{2}\right)^2}}$$

with $j = \exp(\frac{2\mathrm{I}\pi}{3})$, the cubic roots being normalized by

$$\sqrt[3]{-\frac{b}{2} + \sqrt{\left(\frac{a}{3}\right)^3 + \left(\frac{b}{2}\right)^2}}\sqrt[3]{-\frac{b}{2} - \sqrt{\left(\frac{a}{3}\right)^3 + \left(\frac{b}{2}\right)^2}} = -\frac{a}{3}.$$

---

[3] 1465–1526.

A student of Cardan, Ferrari[4], discovered how to reduce degree 4 equations to those of degree 3. We start from the equation

$$x^4 = ax^2 + bx + c,$$

which is equivalent to the following, y being a parameter of the equation,

$$x^4 + 2yx^2 + y^2 = (a + 2y)x^2 + bx + (c + y^2).$$

We look for y such that $(a + 2y)x^2 + bx + (c + y^2)$ is a square $(Ax + B)^2$. In other words, we solve the equation

$$b^2 - 4(a + 2y)(c + y^2) = 0,$$

which is of degree 3 in y. Once we have such a y, all that remains is to solve the equation $x^4 + 2yx^2 + y^2 = (Ax + B)^2$, which is none other than

$$(x^2 + y - Ax - B)(x^2 + y + Ax + B) = 0,$$

or two equations of degree 2!

In all these cases of small degree, the complex roots of the initial general equation are obtained using polynomials in its coefficients as well as various m-th roots of such polynomials: we say that they are expressed by radicals. This is impossible for $n \geq 5$: this is a consequence of the theorem of symmetric functions and Galois theory (cf. Sect. 9.4). This is the most well-known historical success of Galois theory. The theory actually provides much more precise results concerning each individual equation. For example, we can show with the methods developed here that the roots of the equation $X^5 - X - 1$ cannot be expressed by m-th roots of rational numbers!

To conclude this introduction, let us emphasize again that Galois theory is not limited to these applications of historical interest. Indeed, far from it, it has multiple facets, very deep, governing vast aspects of both algebra and number theory and geometry. It is the fine study of the linear representations of the "absolute" Galois group of $\mathbf{Q}$—through notably a particular case of the Langlands conjectures—that allowed Wiles to prove Fermat's theorem. In short, this course is only the *beginning* of a long story, far from being over.

---

[4] 1522–1565.

Chapter 2
Basic Concepts of Group Theory

In this chapter, we recall important elements of group theory that will be useful later on.

2.1 Groups

Definition 2.1.1 A *group* is a set G equipped with an internal composition law, that is, a map $* : G \times G \to G$, such that

- $*$ is associative: $(a * b) * c = a * (b * c)$ for all $a, b, c \in G$,
- $*$ has a neutral element e: $a * e = e * a = a$ for all $a \in G$,
- $*$ has an inverse: for all $a \in G$, there exists an $a^{-1} \in G$ such that $a * a^{-1} = a^{-1} * a = e$.

We say that $*$ is the *law* of the group.

The law $*$ is said to be commutative if $x * y = y * x$ for all x, y in the group. We then say that the group is commutative or abelian. In practice, we often denote the group law additively ($a + b$ instead of $a * b$, the neutral element being denoted by 0) when the group is commutative, and multiplicatively otherwise (ab instead of $a * b$, the neutral element being denoted by 1). However, this is not universal, as we will encounter. For example, the set of integers $\mathbf{Z}$ equipped with the sum is a commutative group while $GL_n(\mathbf{R})$ equipped with the matrix product is a non-commutative group as soon as $n > 1$.

© The Author(s), under exclusive license to Springer Nature Switzerland AG 2024
D. Hernandez, Y. Laszlo, *Introduction to Galois Theory*, Springer Undergraduate
Mathematics Series, https://doi.org/10.1007/978-3-031-66182-2_2

Definition 2.1.2 A *subgroup* G' of a group G is a non-empty subset of G stable under the group law and the inverse, that is, such that $x * y \in G'$ and $x^{-1} \in G'$ for all $x, y \in G'$.

For example, the subgroups of $\mathbf{Z}$ are the $n\mathbf{Z}$ with $n \geq 0$. An intersection of subgroups is again a subgroup.

For a subset A of a group G, the subgroup $\langle A \rangle$ of G generated by A is defined as the smallest subgroup of G containing A, that is, the intersection of all the subgroups of G containing A. We say that A generates G if $\langle A \rangle = G$. A group is said to be cyclic if it can be generated by a one-element set. For example, the group $\mathbf{Z}$ and its subgroups $n\mathbf{Z}$ are cyclic.

A group morphism is a map $\phi : G \to G'$, with G and G' groups, such that $\phi(x * y) = \phi(x) * \phi(y)$ and $\phi(x^{-1}) = (\phi(x))^{-1}$ for all $x, y \in G$. The neutral element of G is then automatically sent to the neutral element of G'.

The kernel $\text{Ker}(\phi) = \{x \in G | \phi(x) = e'\}$ of the group morphism $\phi : G \to G'$ is the inverse image of the neutral element e' of G' by ϕ. It is a subgroup of G (with the additional property of being "normal", as we will see). Moreover, the image $\text{Im}(\phi) = \{\phi(x) | x \in G\}$ of ϕ is a subgroup of G'.

A group isomorphism is a group morphism that is a bijection. Then, automatically, the inverse bijection is also a group morphism. For example, the map $\phi : \mathbf{Z} \to n\mathbf{Z}$ defined by $\phi(m) = nm$ is a group isomorphism.

2.2 Quotient Groups

Let H be a subgroup of G. We define the set of left translates of H

$$G/H = \{gH, g \in G\}.$$

The cardinality $|G/H|$ of G/H, finite or not, is called the index of H in G.

For $g, g' \in G$, if $g' \in gH$, then $g'H = gH$. This implies that G/H is a partition of G, that is, G is the disjoint union of the left translates. Let us finally note that the mapping $h \mapsto gh$ defines a bijection between H and gH.

Consequently, if G is finite, all left translates have the same cardinality $|H|$. We have thus shown that

$$|G| = |G/H||H|.$$

In particular, the cardinality of H divides that of G: this is Lagrange's theorem. We also obtain that $|G/H|$ is finite and equal to $|G|/|H|$. It may happen that G/H is finite while neither G nor H are (we will see examples later).

Let us return to the general case when G is not necessarily finite. We have the canonical surjection

$$\pi : G \rightarrow G/H$$

which sends g to gH.

We would like to put a **group structure** on G/H so that the canonical surjection π is a group morphism. We must therefore have

$$g_1 g_2 H = g_1 H g_2 H, \forall g_1, g_2 \in G.$$

In particular, for $g_1 g_2 = e$, we find that necessarily

$$H = g_1 H g_1^{-1}.$$

Definition 2.2.1 A subgroup H of G is said to be *normal* if

$$gHg^{-1} = H$$

for all $g \in G$. We denote this by $H \triangleleft G$.

For example, if G is abelian, every subgroup of G is normal. However, for example:

Exercise 2.2.2 Show that $GL_n(\mathbf{R})$ is not normal in $GL_n(\mathbf{C})$.

For symmetric groups (see the reminders in the next section), the subgroup S_3 of S_4 is not normal.

In general, we have

Lemma 2.2.3 *The kernel of any group morphism is normal.*

Proof Let $\phi : G \rightarrow G'$ be a group morphism. If $\phi(g) = e'$ is the neutral element of G', then for $h \in G$, we have

$$\phi(hgh^{-1}) = \phi(h)\phi(g)\phi(h)^{-1} = \phi(h)\phi(h)^{-1} = e'$$

and therefore $hgh^{-1} \in \mathrm{Ker}(\phi)$.   □

Another way to express the property $H \triangleleft G$ is to say $gH = Hg$ for all $g \in G$. We then have

$$g_1 g_2 H = (g_1 H)(g_2 H) \forall g_1, g_2.$$

That allow us define, in a unique way, a group structure, called the *quotient group*, on G/H making π a morphism: for g_1H and g_2H in G/H, we set

$$(g_1 H) * (g_2 H) = (g_1 g_2) H.$$

From the above, this definition does not depend on the choice of g_1 and g_2 in their class, but only on the classes g_1H and g_2H themselves. It is easy to verify that we obtain a group,[1] with neutral element H $\in$ G/H and inverse law $(g H)^{-1} = g^{-1} H$.

Note that as H is the kernel of the canonical surjection G $\to$ G/H, every normal subgroup is the kernel of a group morphism, which is a converse of Lemma 2.2.3.

Example 2.2.4 Let $\mathbf{Z}/n\mathbf{Z}$ be the group of integers modulo $n \in \mathbf{Z}$. It is indeed the quotient of the group $\mathbf{Z}$ by the normal subgroup $n\mathbf{Z}$, which justifies the notation. For all $n \in \mathbf{Z}$, the group $\mathbf{Z}/n\mathbf{Z}$ is cyclic. Note that if $n \neq 0$, $\mathbf{Z}$ and $n\mathbf{Z}$ are infinite, while $\mathbf{Z}/n\mathbf{Z}$ is finite of order n. In this latter case, the other generators of $\mathbf{Z}/n\mathbf{Z}$ are the images in $\mathbf{Z}/n\mathbf{Z}$ of the $m \in \{1, \ldots, n-1\}$ that are coprime with n.

Let us return to the general case of an arbitrary group morphism.

Proposition 2.2.5 *Let* $f : G \to G'$ *be a group morphism. Then* $\mathrm{Im}(f)$ *is isomorphic to the quotient group* G/Ker(f).

Proof We have seen that Ker(f) is normal in G, so G/Ker(f) is indeed a group. For $g, g' \in$ G such that $g' \in g$Ker(f), we have $f(g') = f(g)$. So f induces a map

$$\overline{f} : (G/\mathrm{Ker}(f)) \to \mathrm{Im}(f).$$

This is clearly a surjective map that is a group morphism. Moreover, if $(g\mathrm{Ker}(f)) \in \mathrm{Ker}(\overline{f})$, then $f(g) = 1$, and therefore $g\mathrm{Ker}(f) = \mathrm{Ker}(f)$. So $\overline{f}$ is injective and is a group isomorphism.   □

2.3 Supplement on Commutative Groups

Proposition 2.3.1 *Let* G *be a commutative group generated by a finite number of elements. Then* G *is a product of cyclic groups.*

Proof Let us denote by $n(G)$ the minimal number of generators of G. If $n(G) = 0$, the group is zero, if $n(G) = 1$, the group is cyclic (and non-zero). Let us show by induction that G is a product of at most $n(G)$ cyclic groups.

---

[1] The group G/H can also be characterized by a universal property. We will see later, in the case of quotient rings, an example of a universal property.

Assume the theorem has been proven for $n(G) = n \geq 1$ and let G be such that $n(G) = n + 1$. If there exist $n(G)$ generators $\gamma = (\gamma_i)$ of G such that the surjection

$$\underline{\gamma} : \begin{cases} \mathbf{Z}^{n+1} \to & G \\ (n_i) \mapsto \sum_i n_i \gamma_i \end{cases}$$

is injective, then $G \simeq (\mathbf{Z})^{n(G)}$ and the result follows.

Otherwise, the kernels of the morphisms $\underline{\gamma}$ are never zero and we can define

$$|\gamma| = \min \left\{ |n_i| \text{ such that } n_i \neq 0 \text{ and } \underline{\gamma}(n_i) = 0 \right\} \in \mathbf{N} - 0.$$

Let us then choose γ such that the integer $d = |\gamma| \geq 1$ is minimal when γ describes all the families of generators of G of cardinality $n + 1$. If necessary, by reordering the γ_i, $i = 0, \ldots, n$ (and changing a sign), we can therefore assume that we have a relation $d\gamma_0 + \sum_{i>0} n_i \gamma_i = 0$. Note that $d \geq 2$ because otherwise the n generators γ_i, $i \geq 1$, generate G. Let us write the Euclidean division $n_i = dq_i + r_i$, $0 \leq r_i < d$, and set

$$\tilde{\gamma}_0 = \gamma_0 + \sum_{i \geq 1} q_i \gamma_i \text{ and } \tilde{\gamma}_i = \gamma_i, \text{ if } i \geq 1.$$

The $\tilde{\gamma}_i$ generate G and we have $d\tilde{\gamma}_0 + \sum_{i \geq 1} r_i \tilde{\gamma}_i = 0$, which provides a non-zero element of the kernel of $\tilde{\underline{\gamma}}$. By minimality of d, we therefore have $r_i = 0$ for all $i > 0$ and $d\tilde{\gamma}_0 = 0$. By minimality again, the order of $\tilde{\gamma}_0$ is exactly d.

Let us then set $H = \langle \tilde{\gamma}_i, \ i \geq 1 \rangle$. We have $n(H) = n$ by construction and the addition morphism

$$\begin{cases} \mathbf{Z}/d\mathbf{Z} \times H & \to & G \\ (\bar{n}, h) & \mapsto n\tilde{\gamma} + h \end{cases}$$

is surjective. An element $(\bar{n}, h)$ of the kernel satisfies $n\tilde{\gamma} \in H$. If r is the remainder of the division of n by d, we therefore have $r\tilde{\gamma}_0 + h = 0$, which provides an element of the kernel of $\tilde{\underline{\gamma}}$. By minimality again, we have $r = 0$ and therefore $h = 0$, then $n\tilde{\gamma}$ and therefore $\bar{n}$ is equal to 0 because the order of $\tilde{\gamma}$ is d. Thus $G \simeq \mathbf{Z}/d\mathbf{Z} \times H$. □

Remark 2.3.2 The previous result is a weakened form of Kronecker's theorem on the structure of finite type abelian groups (cf. for example [Wae49]). It can be demonstrated in a completely analogous way by Gauss algorithm on matrices with integer coefficients.

Exercise 2.3.3 (Difficult) Show by induction on n that every subgroup G of $\mathbf{Z}^n$ is of the form $\mathbf{Z}^m$, $m \leq n$. [Hint: consider the restriction to G of the projection $(x_i) \mapsto x_n$ then consider a pre-image of a generator of its image as well as a $\mathbf{Z}$-base of the kernel.] Deduce that a subgroup of a finitely generated commutative group is finitely generated. Show in general that if $\mathbf{Z}^n$ and $\mathbf{Z}^m$ are isomorphic, we have

$n = m$. [Hint: consider the matrix defined by such an isomorphism as well as that of its inverse and reduce to usual linear algebra.]

2.4   Exact Sequences

Let G_1, G_2, G_3 be three groups and

$$f_1 : G_1 \to G_2, \; f_2 : G_2 \to G_3$$

two group morphisms.

Definition 2.4.1 We say that the sequence

$$G_1 \xrightarrow{f_1} G_2 \xrightarrow{f_2} G_3$$

is *exact* if $\mathrm{Im}(f_1) = \mathrm{Ker}(f_2)$.

When G_1 (resp. G_3) is reduced to the trivial group $\{1\}$, exactness means that f_2 is injective (resp. f_1 is surjective).

Consider a longer sequence of group morphisms, i.e. $G_1, \dots, G_n$ are groups and $f_i : G_i \to G_{i+1}$ are group morphisms:

$$G_1 \xrightarrow{f_1} G_2 \xrightarrow{f_2} G_3 \xrightarrow{f_3} \cdots \xrightarrow{f_{n-2}} G_{n-1} \xrightarrow{f_{n-1}} G_n.$$

Definition 2.4.2 We say that such a sequence is *exact* if all the sub-sequences with three consecutive terms

$$G_{i-1} \xrightarrow{f_{i-1}} G_i \xrightarrow{f_i} G_{i+1}$$

are exact.

Let G_1, G_2, G_3 be three groups. We have the canonical group morphisms $\{1\} \to G_1$ and $G_3 \to \{1\}$. Let $G_1 \xrightarrow{f_1} G_2$ and $G_2 \xrightarrow{f_2} G_3$ be two group morphisms.

Proposition 2.4.3 *The sequence*

$$\{1\} \to G_1 \xrightarrow{f_1} G_2 \xrightarrow{f_2} G_3 \to \{1\}$$

is exact if and only if

(i) *f_1 is injective,*
(ii) *f_2 is surjective,*
(iii) $\mathrm{Im}(f_1) = \mathrm{Ker}(f_2)$.

In this case, $\mathrm{Im}(f_1) \simeq \mathrm{G}_1$ is a normal subgroup of G_2 and G_3 is isomorphic to the quotient group $\mathrm{G}_2/\mathrm{Im}(f_1)$.

Proof The exactness of the three sub-sequences with three terms of the long sequence is equivalent to properties (i), (ii) and (iii). The first item is therefore clear. Now, suppose that the sequence is indeed exact. As f_1 is injective, we have $\mathrm{Im}(f_1) \simeq \mathrm{G}_1$. As f_2 is surjective, we have (2.2.3)

$$\mathrm{G}_3 \simeq \mathrm{G}_2/\mathrm{Ker}(f_2) \simeq \mathrm{G}_2/\mathrm{Im}(f_1).$$

□

Let us revisit the example of an arbitrary group morphism $f : \mathrm{G} \to \mathrm{G}'$ as in the previous section. We then have a canonically associated exact sequence

$$\{1\} \to \mathrm{Ker}(f) \xrightarrow{i} \mathrm{G} \xrightarrow{f} \mathrm{Im}(f) \to \{1\}, \qquad (2.4.1)$$

where $i : \mathrm{Ker}(f) \to \mathrm{G}$ is the inclusion.

In general, for H a normal subgroup of a group G, we have a canonically associated exact sequence

$$\{1\} \to \mathrm{H} \xrightarrow{i} \mathrm{G} \xrightarrow{\pi} \mathrm{G}/\mathrm{H} \to \{1\}.$$

A "3-term exact sequence"

$$\{1\} \to \mathrm{G}_1 \xrightarrow{f_1} \mathrm{G}_2 \xrightarrow{f_2} \mathrm{G}_3 \to \{1\}$$

is called a *short exact sequence*; we then say that G_2 is an *extension* of G_3 by G_1.

Example 2.4.4 Let D_n be the group of plane isometries leaving invariant a regular polygon with n sides centered at 0 (D_n is called a dihedral group). D_n contains the group of rotations leaving this polygon invariant. This subgroup is cyclic, generated by the rotation of angle $2\pi/n$, and therefore is isomorphic to $\mathbf{Z}/n\mathbf{Z}$. As an isometry preserves distances, an element of D_n is determined by the image of two consecutive vertices. Therefore, D_n has at most $2n$ elements and is generated by rotations and an orthogonal symmetry with respect to a line. We thus obtain a short exact sequence

$$\{1\} \to \mathbf{Z}/n\mathbf{Z} \to \mathrm{D}_n \to \mathbf{Z}/2\mathbf{Z} \to \{1\},$$

and D_n is a non-abelian extension of $\mathbf{Z}/2\mathbf{Z}$ by $\mathbf{Z}/n\mathbf{Z}$.

Exercise 2.4.5 Show that the group

$$\mathrm{GL}_n(\mathbf{C})/\mathrm{SL}_n(\mathbf{C})$$

is isomorphic to $\mathbf{C}^*$.

2.5   Group Actions

One of the main motivations for studying the concept of a group is that groups can act on sets.

Let E be a set and Bij(E) the set of bijections of E. Then Bij(E) is a group with respect to the law of composition, the neutral element being the identity mapping of E and the inverse of a bijection its inverse bijection. It is called the group of permutations of E.

> **Definition 2.5.1** Let E be a set and G a group. An *action* of G on E is the data of a group morphism $\phi : G \to \text{Bij}(E)$.

For example, the identity morphism of Bij(E) defines an action of Bij(E) on E. The group of linear isomorphisms GL(V) of a vector space V naturally acts on this space V, but also on the set $\mathbf{P}(V)$ of vector lines of V, or on the set of sub-vector spaces of V of fixed dimension. The map $\phi : \{1, -1\} \to \text{Bij}(\mathbf{C})$ such that $\phi(1)$ is the identity and $\phi(-1)$ is complex conjugation defines an action of the two-element group $\{1, -1\}$ on $\mathbf{C}$. The map that associates $\theta \in \mathbf{R}$ with the rotation of the plane R_θ of angle θ defines an action of $\mathbf{R}$ on the unit circle of the plane. There are many other group actions, and we will see many examples later. Here is an example that will be particularly useful in the following.

Example 2.5.2 Let G be a group. The map $\phi_g : G \to G$ defined by $\phi_g(h) = ghg^{-1}$ is a group morphism. Furthermore, $\phi : G \xrightarrow{g \mapsto \phi_g} \text{Bij}(G)$ is also a group morphism. We thus obtain the action by conjugation of G on itself. Note that for H a subgroup of G, $\phi_g(H)$ is also a subgroup of G. Thus G acts by conjugation on the set of its subgroups.

In the remainder of this section, E is a set equipped with an action $\phi : G \to \text{Bij}(E)$ of a group G. For $g \in G$ and $x \in E$, the element $(\phi(g))(x)$ will simply be denoted $g \cdot x$.

Definition 2.5.3 Let $x \in E$.
 The *orbit* of x under the action of G is the set $\{g \cdot x | g \in G\} \subset E$.
 The *stabilizer* of x in G is the set $\{g \in G | g \cdot x = x\} \subset G$.

Note that the set of orbits is a partition of E because it is clear that for $x, y \in E$, x is in the orbit of y if and only if y is in the orbit of x.
 Also note that the stabilizer of x is clearly a subgroup of G.
 A subgroup H of a group G is normal in G if and only if its orbit under the action of G by conjugation is {H}.

Definition 2.5.4 A *fixed point* (or *invariant element*) of E under the action of G is an element $x \in E$ whose stabilizer is G.

For H a subgroup of G, the set E^H of invariant elements under H is

$$E^H = \{x \in E | \forall h \in H, h \cdot x = x\} \subset E.$$

To say that a point x is fixed under the action of G is to say that the orbit of x under the action of G is reduced to $\{x\}$, or that $g \cdot x = x$ for all $g \in G$.

The set E^H is nothing other than the set of fixed points of E under the action of H induced by that of G.

Definition 2.5.5 The action of G on E is said to be *faithful* if ϕ is injective.

The action of G on E is said to be transitive if E consists of only one orbit.

To say that the action is transitive is to say that for all $x, y \in E$, there exists a $g \in G$ such that $g \cdot x = y$.

Exercise 2.5.6 Let V be a finite-dimensional vector space of dimension n over a field k. For $0 \le d \le n$, show that the action of GL(V) on the set of subspaces of V of dimension d is transitive.

Definition 2.5.7 A subset $F \subset E$ is said to be *globally invariant* under the action of G if for all $g \in G$, we have $g(F) \subset F$.

A subset formed of invariant elements is globally invariant, but the converse is generally false.

2.6 Symmetric Groups

Let $n \ge 2$ be an integer.

Definition 2.6.1 The *symmetric group* is $S_n = \text{Bij}(X)$ where $X = \{1, \ldots, n\}$.

The symmetric group S_n naturally acts on X. It has cardinality $n!$. The group S_1 being the trivial group, we assume in the following that $n \ge 2$. If $\sigma \in S_n$, we define an equivalence relation by saying that two elements $x, y \in X$ are equivalent if and only if there exists a $j \in \mathbf{Z}$ such that $y = \sigma^j(x)$. An equivalence class is also called a σ-orbit.[2] As with any equivalence relation, X is a disjoint union of σ-orbits $O_i(\sigma)$. Let

$$n_1 \ge n_2 \ge \cdots \ge n_N$$

---

[2] This is an orbit under the action of the subgroup of S_n generated by σ.

be the (possibly empty) ordered sequence of the cardinalities of the orbits not reduced to one element. We thus have $N = 0$ if and only if $\sigma = \text{Id}$.

Definition 2.6.2 The *type* of σ is the N-tuple $\bar{n} = (n_1, \ldots, n_N)$.
    We say that σ is a cycle (of length $d = n_1$) if $N = 1$: we then speak of a d-cycle. The support of a cycle is its unique orbit not reduced to one element.

The length of a cycle is the cardinality of its support. Cycles of length 2 are called transpositions. They generate the group S_n.

A cycle of length $d > 1$ may be (non uniquely) written in the form

$$\sigma = (x, \sigma(x), \ldots, \sigma^{d-1}(x))$$

where x is an arbitrary element in the non-trivial orbit of σ. For example, the cycle of length 3 denoted $(3, 7, 5)$ fixes any element distinct from $3, 7, 5$ and cyclically permutes the other elements as in the diagram

$$3 \to 7 \to 5 \to 3.$$

Two cycles commute if and only if their supports are disjoint. The product of such cycles is therefore well defined, the order not intervening. We then have the following property.

Proposition 2.6.3 *Every permutation can be uniquely written as a product (possibly empty) of cycles with disjoint supports.*

For a cycle $(a_1, \ldots, a_m) \in S_n$ $(m \leq n)$ and $\sigma \in S_n$, we have the formula

$$\sigma \circ (a_1, \ldots, a_m) \circ \sigma^{-1} = (\sigma(a_1), \ldots, \sigma(a_m)). \tag{2.6.1}$$

This ensures that the conjugate of a cycle is a cycle of the same length and that moreover two cycles are conjugate if and only if they have the same length. More generally, we have the following characterization, an immediate corollary of this remark and the previous proposition.

Proposition 2.6.4 *Two permutations are conjugate if and only if they have the same type.*

The reader is strongly advised to do the following exercise.

Exercise 2.6.5 Show that the transpositions $(i, i + 1)$, $1 \leq i \leq n - 1$, generate S_n. Deduce that $(1, \ldots, n)$ and $(1, 2)$ generate S_n. Show that S_n is generated by any triple (a, b, c) with a, b, c cycles of length $n, n - 1$ and 2 respectively.

We define the signature

$$\epsilon : S_n \to \{1, -1\}$$

by the formula

$$\epsilon(\sigma) = (-1)^{\sum_{i=1}^{N}(n_i - 1)},$$

where σ is of type $(n_1, \ldots, n_N)$.
We verify that the signature is also $(-1)^i$ where

$$i = \text{card}\{(x, y) \in X^2 \text{ such that } x > y \text{ and } \sigma(x) < \sigma(y)\}$$

is the *number of inversions* of σ.
We have the following fundamental result.

Proposition 2.6.6 *The signature ϵ is the unique surjective group morphism $S_n \to \{1, -1\}$.*

Proof Using the formula with the number of inversions, we indeed obtain that the signature is a group morphism. The signature of a transposition being equal to -1, the signature is surjective. Now, the transpositions generate S_n. A group morphism $\phi : S_n \to \{1, -1\}$ is therefore uniquely determined by its value on transpositions. All transpositions being conjugate, ϕ takes the same value on all transpositions. If this value is -1, ϕ is the signature. Otherwise, ϕ is the trivial morphism, not surjective. $\square$

The kernel A_n of the signature is therefore a normal subgroup, of cardinality $n!/2$: it is called the alternating group. We have an exact sequence

$$\{1\} \to A_n \to S_n \to \{1, -1\} \to \{1\}.$$

Transpositions have a signature of -1. We deduce that the signature of a permutation σ is $(-1)^N$, where N is the number of transpositions involved in *one* decomposition of σ into a product of transpositions.

Exercise 2.6.7 Show that A_n is generated by the 3-cycles as soon as $n \geq 3$.

Exercise 2.6.8 Let H be a subgroup of finite index in G. Show that if the index of H in G is 2, then H is normal in G and the quotient G/H is canonically isomorphic to $\{\pm 1\}$. Let n be an integer ≥ 2. Show that the alternating group A_n is the unique subgroup of S_n of index 2.

2.7 Solvable Groups

The class of commutative groups is not stable under short exact sequences. We need a larger class: that of solvable groups.

Definition 2.7.1 A group G is said to be *solvable* if it has a decreasing sequence of subgroups

$$\{1\} = G_n \subset \cdots \subset G_0 = G$$

such that for $0 \leq i \leq n-1$, the group $G_{i+1} \subset G_i$ is normal in G_i and the quotient group G_i/G_{i+1} is commutative.

We will simply say that "the successive quotients are commutative".

Note that a commutative group is solvable, just set $G_1 = \{1\}$. Also note that a non-commutative solvable group always contains a non-trivial normal subgroup, the subgroup G_1.

Exercise 2.7.2 Show that the group of complex matrices of size $n \geq 2$ with determinant 1 is not solvable.

Definition 2.7.3 The *derived subgroup* DG of a group G is the subgroup of G generated by all the commutators $[a, b] = aba^{-1}b^{-1}$ with $a, b \in G$.

Note that DG is usually not equal to the subset $\{[a, b]|a, b \in G\}$, which is not a subgroup in general.

Lemma 2.7.4 DG *is a normal subgroup of G, and the quotient group G/DG is commutative.*

Proof Let $g \in G$. The map ϕ_g of Example 2.5.2 is a group morphism. Consequently, for $a, b \in G$, we obtain $\phi_g([a, b]) = [\phi_g(a), \phi_g(b)] \in DG$. Therefore DG is normal. The quotient group is commutative by construction. $\qquad\square$

We then define by induction on n the sequence of subgroups $(D^n G)_{n \geq 0}$

$$D^0 G = G \text{ and } D^{n+1} G = D D^n G \text{ if } n \geq 0.$$

Lemma 2.7.5 *G is solvable if and only if $D^n G$ is trivial for n large enough.*

Proof If G is solvable and G_i is as in the definition, the image of a commutator in the abelian group G_0/G_1 is trivial so that $D^1 G$ is contained in G_1. By induction, we show that $D^i G$ is contained in G_i and therefore $D^n G$ is trivial. Conversely, if $D^n G$ is trivial, we set $G_i = D^i G$.   □

Proposition 2.7.6 *If*

$$1 \rightarrow G_1 \rightarrow G_2 \rightarrow G_3 \rightarrow 1$$

is exact, then G_2 is solvable if and only if G_1 and G_3 are solvable.

Proof On the one hand, we have $D^n G_2 \rightarrow D^n G_3$ surjective and $D^n G_1 \rightarrow D^n G_2$ injective so that G_2 being solvable implies G_1 and G_3 are solvable. Conversely, if $D^n G_3$ is trivial, the image of $D^n G_2$ in G_3 is zero and therefore $D^n G_2$ is contained in G_1. If now we also have $D^m G_1 = 1$, we deduce $D^{m+n} G_2 \subset D^m G_1 = 1$, hence the converse.   □

In fact, we have better: the class of solvable groups is stable under extension, which given the above, is written

Corollary 2.7.7 *If G has an increasing sequence of subgroups*

$$1 = G_0 \subset \cdots \subset G_n = G$$

with G_i normal in G_{i+1} and G_{i+1}/G_i solvable, then G is solvable.

Exercise 2.7.8 We aim to show that the group B of matrices of $GL_n(k)$ that are upper triangular is solvable (k is a field). Let U be the subgroup of B of matrices whose eigenvalues are all equal to 1 (unipotent matrices).

(1) Show that we have an exact sequence of groups

$$1 \rightarrow U \rightarrow B \rightarrow (k^*)^n \rightarrow 1.$$

Deduce that B is solvable if and only if U is solvable.

Let (e_i) be the canonical basis of k^n. For $i \leq n$, let F_i be the subspace of k^n generated by $e_1, \ldots, e_i$. We have $F_i = (0)$ if $i \leq 0$ and $F_n = k^n$. For all $f \in U$, we denote by $\ln(f)$ the matrix $f - \mathrm{Id}$. For all $j = 0, \ldots, n$, let U_j be the subset of U comprising the matrices f such that $\ln(f)(F_i) \subset F_{i-j}$ for $i \leq n$.

(2) Verify that we have

$$(1) = U_n \subset U_{n-1} \subset \cdots \subset U_1 = U.$$

Show that U_i is a normal subgroup of U for all $i \leq n$ and therefore also of U_{i-1}.
(3) Let $f \in U_j$. Show that for all $i \leq n$, the restriction $\ln(f)_{i,j}$ of $\ln(f)$ to F_i induces a linear map of F_i/F_{i-j-1} which is zero if and only if $\ln(f)(F_i) \subset F_{i-j-1}$.
(4) Show that the map

$$\ln_j : \begin{cases} U_i \to \prod_i \mathrm{End}(F_i/F_{i-j}) \\ f \mapsto \quad (\ln(f)_{i,j}) \end{cases}$$

is a group morphism and calculate its kernel.
(5) Deduce that U is solvable. Conclude.

Lemma 2.7.9 *The groups S_3 and S_4 are non-commutative and solvable.*

Proof The formula $(1, 2)(2, 3) = (1, 2, 3) \neq (1, 3, 2) = (1, 3)(2, 3)$ shows that S_3 and S_4 are not commutative. Now, for $n \geq 2$, the exact sequence

$$1 \to A_n \to S_n \to \{\pm 1\} \to 1$$

and Proposition 2.7.6 ensure that S_n is solvable if and only if A_n is. Since A_3 is cyclic of order 3, it is solvable. For A_4, we can observe that

$$K = \{\mathrm{Id}, (12)(34), (13)(24), (14)(23)\}$$

is a normal subgroup of A_4 (and of S_4 for that matter; it is called the Klein group). Since A_4 has cardinality 12, the quotient has cardinality 3, and therefore is cyclic like any group of prime order. The result follows again thanks to Proposition 2.7.6.
□

The following result is fundamental for the sequel.

Proposition 2.7.10 *If $n \geq 5$, we have $D(A_n) = A_n$ and therefore S_n is not solvable.*

Proof $n \geq 3$, so A_n is generated by the 3-cycles (cf. Exercise 2.6.5). Let $\gamma = (a, b, c)$ be a 3-cycle. We have $\gamma^2 = (a, c, b)$. Let $\sigma \in S_n$ send the triplet (a, b, c) to (a, c, b), that is, such that $\sigma(a, b, c)\sigma^{-1} = (a, c, b)$. Let d, e be distinct from

a, b, c (this is possible because $n \geq 5$). We also have

$$\sigma \circ (d, e)(a, b, c) = (a, c, b)$$

and therefore we can assume, if necessary changing σ to $\sigma \circ (d, e)$, that σ is in A_n. However

$$\sigma \circ \gamma \circ \sigma^{-1} = \gamma^2$$

and therefore $\gamma \in D(A_n)$.

For the last point, we observe that $D(S_n) \subset A_n$.   □

Exercise 2.7.11 Let

$$X = \{(1, 2)(3, 4), (1, 3)(2, 4), (1, 4)(2, 3)\}$$

be the set of permutations of S_4 of type $(2, 2)$. Show that S_4 acts on X by conjugation. Deduce that there exists a morphism $\pi : S_4 \to S_3$ and calculate its kernel. Show that π is surjective and deduce that S_4 is solvable. This proves the result of Exercise 2.7.2.

Exercise 2.7.12 (Difficult) Let G be a group of cardinality p^n with p prime.[3] We propose to show by induction on n the existence of an increasing sequence of subgroups G_i, $i = 1, \ldots, n$, of G of cardinality p^i with G_i normal in G_{i+1}. This shows in particular that G is solvable.

(a) Let H be a normal subgroup of G. Show that if the statement is true for H and G/H it is true for G.
(b) Treat the commutative case.
(c) By making G operate on itself by conjugation, show that the center of G is not reduced to 1. Conclude.

---

[3] The interest in these groups is partly due to the fact that if p^n is the maximum power of p dividing the cardinality of a finite group G, then G has a subgroup of order p^n (called a Sylow p-group) and that all these subgroups are conjugate (see for example [Bou89]).

Chapter 3
Basic Concepts of Ring Theory

As in the previous chapter, we recall here some important elements of ring theory.

3.1 Rings

Let us recall some fundamental definitions.

Definition 3.1.1 A *ring* is a set A equipped with two commutative operations

$$+ : A \times A \to A \text{ and } \times : A \times A \to A$$

such that $(A, +)$ is a group and $\times$ is associative, equipped with a neutral element, and distributive over $+$:

$$x \times (y + z) = x \times y + x \times z \text{ for all } x, y, z \in A.$$

In other words, a ring is a set equipped with an addition and a multiplication that allows us to do calculations as we would, for example, on the integers or real numbers, except that one cannot generally divide by a non-zero element. Sometimes we will simply write $x \cdot y$ or xy for $x \times y$. The neutral element of $+$ is denoted 0, and the neutral element of $\times$ is denoted 1.

We note that, in view of our definition above, by a 'ring' we mean a commutative unitary ring.

© The Author(s), under exclusive license to Springer Nature Switzerland AG 2024   25
D. Hernandez, Y. Laszlo, *Introduction to Galois Theory*, Springer Undergraduate
Mathematics Series, https://doi.org/10.1007/978-3-031-66182-2_3

Definition 3.1.2 A *field* is a non-zero ring in which every non-zero element has an inverse with respect to ×.

Example 3.1.3 The set of integers $\mathbf{Z}$, integers modulo n (denoted $\mathbf{Z}/n\mathbf{Z}$), real-valued functions on a set, and convergent power series (equipped with the usual laws) are examples of rings, but not generally fields. The set $\mathbf{Z}/p\mathbf{Z}$ of integers modulo a prime p is a field, like the sets $\mathbf{Q}$, $\mathbf{R}$ and $\mathbf{C}$ of rational, real and complex numbers (equipped with the usual laws).

Generally, if not specified and the context is clear, the letter A will designate a ring while k will designate a field.

Definition 3.1.4 A *morphism* of rings $f : A \rightarrow B$ is a map such that $f(1) = 1$ and which satisfies

$$f(a+b) = f(a) + f(b) \text{ and } f(ab) = f(a)f(b)$$

for all $a, b \in A$. Its *kernel* $\text{Ker}(f) \subset A$ is the set of elements sent to zero by f. The set of these morphisms is denoted $\text{Hom}(A, B)$.

Note that necessarily $f(0) = 0$ (uniqueness of the neutral element in a group) and $f(-a) = -f(a)$ for all $a \in A$.

Definition 3.1.5 An *ideal* I of a ring A is a subgroup of A with respect to $+$ such that $xy \in I$ for all $x \in A$, $y \in I$.

This notion is different from the notion of sub-ring, for which the last condition is replaced by stability under ×.

Every ideal I of $\mathbf{Z}$ is generated by an element, that is to say, I is of the form $I = n\mathbf{Z}$ with $n \in \mathbf{Z}$. This is a well-known consequence of Euclidean division in $\mathbf{Z}$. As we will see in the next section, this is also the case for the ring of polynomials over a field.

The kernel of a ring morphism is always an ideal.

Exercise 3.1.6 Show that the pre-image of a subgroup of $\mathbf{Z}/n\mathbf{Z}$ under the canonical projection is a subgroup of $\mathbf{Z}$ containing $n\mathbf{Z}$. Deduce that the subgroups of $\mathbf{Z}/n\mathbf{Z}$ are cyclic of cardinality $d|n$, generated by the class of $\frac{n}{d}$. In particular, the map that associates to a subgroup of $\mathbf{Z}/n\mathbf{Z}$ its cardinality is a bijection onto the set of (positive) divisors of n.

3.2 Rings of Polynomials

Let A be a ring. Then consider the set A[X] of formal polynomials with one variable with coefficients in A, in other words, sequences of elements of A all but finitely many of which are zero. Equipped with the usual addition and multiplication of polynomials, A[X] has a natural ring structure.

   When $A = k$ is a field, we have Euclidean division in $k[X]$: for all A, B $\in k[X]$ with B $\neq 0$, there exists a unique pair (Q, R) of polynomials such that

$$A = QB + R \text{ and } \deg(R) < \deg(B).$$

As a consequence, we obtain, as for the ring of integers, that all ideals of $k[X]$ can be generated by a single element, that is, they are of the form $I = P(X)k[X]$ with $P(X) \in k[X]$. Such an ideal is then denoted (P).

   This allows us to define the GCD (resp. the LCM) of two polynomials P, Q $\in k[X]$, at least one of them being non-zero, as the generator of the ideal (P) + (Q) (resp. (P) ∩ (Q)). Note that the GCD and the LCM are *a priori* defined up to a multiplicative constant in k^*. However, we usually normalize them by taking the unit generators of the corresponding ideals, that is, by dividing by the leading coefficient, which has the virtue of making them well defined as polynomials and not just up to an invertible multiplicative constant.

Lemma 3.2.1 *The GCD (resp. LCM) of P, Q is the Greatest Common Divisor (resp. the Least Common Multiple) of P, Q with respect to the divisibility ordering.*

Proof Let us check the result for $\Delta = $ GCD(P, Q), the proof for the LCM being entirely analogous. Because P, Q $\in$ (P) + (Q) = (Δ) $\neq 0$, one has $\Delta \neq 0$ and Δ is a divisor of both P and Q. Assume D is a (non-zero) common divisor of P, Q. Then (P) $\subset$ (D) and (Q) $\subset$ (D) and therefore (Δ) = (P) + (Q) $\subset$ (D), which proves D|Δ.   □

   When two non-zero polynomials P and Q are coprime, that is, their GCD equals 1, then we have Bézout's identity, that is, there exist U, V $\in k[X]$ such that

$$PU + QV = 1.$$

The Euclidean algorithm allows us to calculate the GCD of two polynomials P, Q $\in k$. This algorithm depends only on the coefficients of P and Q, and not on the field k. In particular, if L is a field containing k, the polynomials P and Q, viewed as elements of L[X], have the same GCD as when they are viewed as elements of $k[X]$. We say that the GCD is invariant under field change.

3.3   Field Morphisms

Note that an ideal I of a ring A is equal to A if and only if it contains 1. This is
equivalent to the fact that I contains an invertible element of A (that is, an element
of A that has an inverse with respect to $\times$). As a consequence:

Proposition 3.3.1 *The only non-zero ideal of a field is the field itself.*

A field morphism is a ring morphism between two fields. In particular, since a
field morphism sends 1 to 1 (non-zero because a field is non-zero), the kernel of a
field morphism is always zero. Thus, we have the following remarkable result.

Proposition 3.3.2 *A field morphism is always injective.*

Thus, a field morphism $\sigma : k \rightarrow k'$ allows us to identify the subfield $\sigma(k)$ of k'
with k (they are isomorphic). We also say that σ defines an embedding of k into k'.
Note that k' is an extension of $\sigma(k)$, in the following sense:

Definition 3.3.3 An *extension* of a field K is a field K' containing K.

Thus, for $\sigma : k \rightarrow k'$ a field morphism, k' can be seen as an extension of k.

When we have three fields $K \subset K' \subset K''$, we say that the extension K'/K is a
sub-extension of the extension K''/K.

3.4   Quotient Rings

Let us recall the construction of the quotient $\overline{A} = A/I$ of a ring A by an ideal I and
especially its properties[1] (compare with Sect. 2.2). The idea is to create a new ring
$\overline{A}$ in which we have "killed" the elements of I. We simply adapt the construction of
$\mathbf{Z}/n\mathbf{Z}$, which will be a particular case of the general construction for $A = \mathbf{Z}$ and
$I = n\mathbf{Z}$.

For $a \in A$, the translate of a is the set

$$\overline{a} = a + I \overset{\text{def}}{=} \{a + i, i \in I\} \subset A.$$

Note that two such translates $(a + I)$ and $(a' + I)$ are equal if and only if $a - a' \in I$.

---

[1] As often in mathematics, the construction is not the most important thing; the properties matter
much more. For example, we know very well how to work on the real field by knowing the
properties of its order without necessarily remembering or even knowing any of its constructions!

The quotient set A/I is the set of translates. The group $(A, +)$ being abelian, I is normal in A. We have seen that then A/I is equipped with a group structure such that

$$(a + I) + (a' + I) \overset{\text{def}}{=} (a + I + a' + I) = (a + a') + I.$$

We observe that A/I is a commutative group with neutral element $\bar{0}$.

Similarly, A/I is equipped with a product defined by

$$(a + I).(b + I) \overset{\text{def}}{=} (a + I)(b + I) + I = ab + aI + bI + I^2 + I = ab + I,$$

where I^2 denotes the ideal generated by the products ii' with $i, i' \in I$. The reader will easily verify the following result.

Proposition 3.4.1 A/I *equipped with the laws* $+$ *and* $\times$ *is a ring. The neutral element of* $+$ *is* $\bar{0}$ *and the neutral element of* $\times$ *is* $\bar{1}$.

Let us define the **canonical surjection**

$$\pi : A \twoheadrightarrow A/I$$

by $a \mapsto \bar{a}$ (the symbol $\twoheadrightarrow$ means that the map is surjective). We regard $\bar{a} = \pi(a)$ as the class A modulo I, exactly as in usual arithmetic.

Proposition 3.4.2 π *is a surjective ring morphism with kernel* $\text{Ker}(\pi) = I$.

Proof We have, for $a, a' \in A$,

$$\pi(1) = \bar{1}, \ \bar{a} \cdot \bar{a'} = \overline{a \cdot a'}, \ \overline{a + a'} = \bar{a} + \bar{a'},$$

so π is a ring morphism. It is clearly surjective. For $a \in \text{Ker}(\pi)$, we have $\bar{a} = \bar{0}$, that is, $a \in I$. Conversely, each $a \in I$ satisfies $\bar{a} = I$.   □

Example 3.4.3 For k a field, we have the ring $k[X]/(P)$, where (P) is the ideal $Pk[X]$ of $k[X]$ generated by a polynomial $P \in k[X]$.

The following statement, called the *universal property of the quotient*, is now clear, and... fundamental.

Starting from a diagram

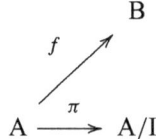

with f a ring morphism such that $f(\mathrm{I}) = 0$, there exists a unique ring morphism $\bar{f}$ making the diagram *commute*

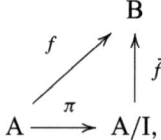

that is, such that $f = \bar{f} \circ \pi$. We also say that f *factors* through π. In set-theoretic terms, this is equivalent to the following theorem, known as the universal property of the quotient.

Let A, B be two rings and I an ideal of A. We define

$$\pi^* : \mathrm{Hom}(A/I, B) \to \{ f \in \mathrm{Hom}(A, B) \mid f(\mathrm{I}) = 0 \}$$

by $\pi^*(g) = g \circ \pi$.

Theorem 3.4.4 (Universal Property of the Quotient) *The map π^* is a bijection.*

We will identify these two spaces without further caution.

Proof Let ϕ, ϕ' be such that $\phi \circ \pi = \phi' \circ \pi$, that is, $(\phi - \phi') \circ \pi = 0$. Since π is surjective, $\phi - \phi'$ is null on $\pi(A) = A/I$ and therefore is zero, hence the injectivity.

Let us move on to surjectivity. Suppose $f \in \mathrm{Hom}(A, M)$ nullifies I and let us look for a pre-image ϕ. Let $t \in A/I$: it is the class of an element a, well determined up to the addition of any element $i \in \mathrm{I}$. Since $f(\mathrm{I}) = 0$, the images under f of all the elements a representing t are one and the same element, which we call $\phi(t)$. By construction, $\phi \circ \pi = f$ and ϕ is obviously a morphism (for example, if $t = \pi(a), t' = \pi(a')$ with $a, a' \in A$, we have

$$\phi(tt') = \phi(\pi(a)\pi(a')) = \phi(\pi(aa')) = f(aa') = f(a)f(a')$$
$$= \phi(\pi(a))\phi(\pi(a')) = \phi(t)\phi(t'),$$

which proves the multiplicativity since ϕ is surjective; additivity is treated in the same way). $\qquad\square$

Remark 3.4.5 If $f : A \to B$ is a ring morphism, we therefore have a canonical factorization $\bar{f} : A/\mathrm{Ker}(f) \to B$ of f through $A \to A/\mathrm{Ker}(f)$ since $f(\mathrm{Ker}(f)) = \{0\}$. As we have precisely "killed" the kernel of f, that of $\bar{f}$ is null so that $\bar{f}$ is *injective*. If f is assumed surjective, we therefore have a canonical isomorphism $\bar{f} : A/\mathrm{Ker}(f) \xrightarrow{\sim} B$.

Let us now prove the following easy but very useful lemma.

Lemma 3.4.6 *The map that associates to an ideal $\bar{J}$ of A/I its pre-image $J = \pi^{-1}(\bar{J})$ identifies the ideals of A/I with the ideals of A containing I. Moreover, the morphism $A \to \bar{A} \to \bar{A}/\bar{J}$ passes to the quotient and induces an isomorphism $A/J \xrightarrow{\sim} \bar{A}/\bar{J}$.*

Proof Since I contains 0, the ideal $\pi^{-1}(I)$ contains $I = \pi^{-1}(0)$. Conversely, if J is an ideal of A containing I, we verify that $\pi(J)$ is an ideal of A/I. The two constructions are clearly the inverse of each other. Moreover, the kernel of the surjection $A \to \bar{A}/\bar{J}$ is the set of $a \in A$ such that $\pi(a) \in \bar{J}$, that is J. By the universal property of the quotient, we have a factorization $A/J \to \bar{A}/\bar{J}$ which obviously remains surjective, but which in addition is injective according to the previous remark. ☐

3.5 The Characteristic

Let us recall the following result:

Lemma 3.5.1 *Let A be a ring. There exists a unique ring morphism $\gamma : \mathbf{Z} \to A$.*

Proof Such a morphism γ satisfies, for $n \geq 0$, $\gamma(n) = 1 + \cdots + 1$ (n times), and for $n \leq 0$, $\gamma(n) = -1 - 1 - \cdots - 1$ (($-n$) times). It is therefore unique. Moreover, these formulas define a ring morphism. ☐

The kernel of γ is an ideal of $\mathbf{Z}$. Therefore, there exists a unique integer $n \geq 0$ such that $\mathrm{Ker}(\gamma) = n\mathbf{Z}$.

Definition 3.5.2 The unique integer $n \geq 0$ such that $\mathrm{Ker}(\gamma) = n\mathbf{Z}$ is called the *characteristic* of A.

For example,[2] the characteristic of $\mathbf{Z}$ is zero and the characteristic of $\mathbf{Z}/n\mathbf{Z}$ is n.

Proposition 3.5.3 *The characteristic of a field k is either zero or a prime number.*

Proof Suppose that the characteristic n of k factorizes as $n = pq$ with p, q positive integers greater than 2. We then have $\gamma(p)\gamma(q) = 0$ in k. Since k is a field, this implies $\gamma(p) = 0$ or $\gamma(q) = 0$. This is a contradiction, because $2 \leq p, q < n$ so p and q are not in $\mathrm{Ker}(\gamma) = n\mathbf{Z}$. □

For example, $\mathbf{Q}$ has characteristic zero while $\mathbf{Z}/p\mathbf{Z}$ has characteristic p (p prime).

Proposition 3.5.4 *Let k be a field. If the characteristic of k is zero, then k is infinite and contains a (unique) subfield isomorphic to $\mathbf{Q}$. If the characteristic of k is a prime number p, then k contains a (unique) subfield isomorphic to $\mathbf{Z}/p\mathbf{Z}$, called the prime subfield of k.*

Proof Consider the unique ring morphism $\gamma : \mathbf{Z} \to k$. Then γ factors through its kernel $n\mathbf{Z}$ to define a canonical injection $\mathbf{Z}/n\mathbf{Z} \hookrightarrow k$. If $n = 0$, then k contains $\mathrm{Im}(\gamma) \simeq \mathbf{Z}$. The subfield of k generated by $\mathrm{Im}(\gamma)$ is then isomorphic to $\mathbf{Q}$. Moreover, k is infinite because it contains the infinite $\mathrm{Im}(\gamma)$. If $n = p$ is a prime number, then k contains $\mathrm{Im}(\gamma) \simeq \mathbf{Z}/p\mathbf{Z}$.

The uniqueness follows since such a subfield is generated by all $m \cdot 1_k$, $m \in \mathbf{Z}$. □

3.6 Domains and Properties of Ideals

Definition 3.6.1 We say that a ring A is a *domain* if it is non-zero and if the product of two non-zero elements of A is non-zero.

---

[2] Some authors define the characteristic only for domains, cf. Sect. 3.6.

Example 3.6.2 The ring $\mathbf{Z}$ is a domain. A field k is a domain. Moreover, $k[X]$ is also a domain. Indeed, the product of two non-zero polynomials $P_1(X) = a_n X^n + \cdots + a_0$ and $P_2(X) = b_m X^m + \cdots + b_0$ is non-zero because it is of degree $n + m$ with leading coefficient $a_n b_m \neq 0$.

A domain A can be embedded in a field K, in the sense that there exists an injective ring morphism from A to K. An important example is the *field of fractions* of A. Its construction is modeled on the construction of $\mathbf{Q}$, which is the field of fractions of $\mathbf{Z}$: we consider the set Frac(A) of equivalence classes of A $\times$ A* for the relation $((a, b)$ equivalent to (c, d) if $ad = bc)$: here (a, b) and (c, d) respectively represent the fractions a/b and c/d. Using the usual rules of addition and multiplication of fractions, we then equip Frac(A) with a field structure. We denote by $(a_s, s \in S)$ the ideal generated by the family of $a_s, s \in S$. If I, J are ideals, we denote by IJ the ideal generated by the products $ij, i \in I, j \in J$. We then speak, abusively, of the *product ideal* of I and J.

Definition 3.6.3 Let I be an ideal of a ring A. We assume $I \neq A$.

- We say that I is *prime* if A/I is a domain.
- We say that I is *maximal* if A/I is a field.

In particular, a field being a domain (Sect. 3.6), a maximal ideal is necessarily prime. Note that the ideal A is neither prime nor maximal.

Lemma 3.6.4 *The inverse image of a prime ideal under a ring morphism is a prime ideal.*

Proof If $\mathfrak{p}$ is prime in A and $f : B \to A$ is a ring morphism, f induces a morphism $B/f^{-1}(\mathfrak{p}) \to A/\mathfrak{p}$ (Theorem 3.4.4), injective by construction. This ensures that $B/f^{-1}(\mathfrak{p})$ is a domain as a subring of a domain and therefore that $f^{-1}(\mathfrak{p})$ is prime. $\qquad \square$

In general, the inverse image of a maximal ideal is not a maximal ideal. For example, the inverse image of the maximal ideal (0) of $\mathbf{Q}$ under the injection $\mathbf{Z} \to \mathbf{Q}$ is null. However, (0) is not maximal in $\mathbf{Z}$ because $\mathbf{Z}/(0) = \mathbf{Z}$ is not a field.

The set of prime ideals of A is denoted Spec(A) and is called the spectrum of A. It is one of the fundamental objects of algebraic geometry.

Definition 3.6.5 An element a of a domain is said to be *irreducible* if it is neither null nor invertible and if its divisors are either invertible or multiples of a.

For example, the irreducibles of **Z** are, up to sign, the prime numbers. The irreducibles of $k[X]$ are the polynomials irreducible in the usual sense of the term.

An ideal of a ring A is said to be proper if it is not equal to A.

> **Proposition 3.6.6** *A proper ideal of a ring* A *is maximal if and only if the only ideal that strictly contains it is* A.

Proof Let I be a proper ideal of A that is maximal. Suppose that J is an ideal of A that contains I strictly. Then there exists an $a \in J \setminus I$. Since A/I is a field, there exists $a, b \in A$ such that $ab \in 1 + I$. Then $1 \in J$ and $J = A$. Conversely, suppose that the only ideal that contains I strictly is A. Let $a \in A \setminus I$. Then the ideal $I + Aa$ generated by I and a is A, so there exists a $b \in A$ such that $1 = ba + i$ with $i \in I$. Then in A/I, $\bar{1} = \bar{b}\bar{a}$ and therefore $\bar{a}$ is invertible in A/I. Therefore A/I is a field and I is maximal.   □

Example 3.6.7 Let k be a field and $P \in k[X]$. Then the ring $k[X]/(P)$ is a field if and only if P is a non-zero irreducible polynomial. For example, for $k = \mathbf{R}$ and the irreducible polynomial $P(X) = X^2 + 1$ in $\mathbf{R}[X]$, we obtain the field

$$\mathbf{R}[X]/(X^2 + 1) \approx \mathbf{C}$$

of complex numbers.

Definition 3.6.8 A domain A such that every ideal of A is generated by a single element is said to be a *principal ring*.

We have seen above that **Z** is principal and that for k a field, the ring $k[X]$ is principal.

Lemma 3.6.9 *Let* A *be a principal ring and* a *a non-zero element of* A. *The following three properties are equivalent:*

(1) a *is irreducible;*
(2) $(a) = aA$ *is prime;*
(3) $(a) = aA$ *is maximal.*

Proof Let us show the equivalence of the properties. The implication (3) $\Rightarrow$ (2) is already known. Suppose (2). Then write a as a product bc. We then have $\bar{b}\bar{c} = \bar{0}$ in $A/(a)$. Since $A/(a)$ is a domain, we have $\bar{b} = \bar{0}$ or $\bar{c} = \bar{0}$, that is $b \in (a)$ or $c \in (a)$. We therefore obtain (1). Finally suppose that (1) is true. Let J be an ideal of A that strictly contains (a). Since A is principal, we have $J = (b)$ for some $b \in A$. Then b divides a, and as $b \notin (a)$, b is invertible and $(b) = I$. This implies (3).   □

For example, as $\mathbf{Z}$ is principal, we immediately obtain:

Proposition 3.6.10 *Let $n > 0$ be an integer. Then the ring $\mathbf{Z}/n\mathbf{Z}$ is a field if and only if n is prime. This also equates to $\mathbf{Z}/n\mathbf{Z}$ being a domain.*

Exercise 3.6.11 Let I be an ideal of A such that for every $i \in$ I, there exists an integer $n \geq 1$ such that $i^n = 0$. Show that the canonical surjection A $\to$ A/I induces a surjection at the level of the group of invertibles. Show that this is false without the condition on I.

3.7   The Rank of a Finite Type Free Module

Let A be a non-zero ring. An A-module V is an abelian group $(V, +)$ equipped with a law $A \times V \to V$ satisfying the axioms of a vector space (in particular, if A is a field, an A-module is simply a vector space over k). We define in the same way the notion of isomorphism between A-modules, of free part and of generating part. A finite type A-module is a module that admits a finite generating part.

A free A-module (of finite type) is a module isomorphic to A^n with $n \geq 0$ an integer. The question is whether the n in question is unique. In other words, does the existence of an isomorphism $A^n \approx A^m$ imply $n = m$?[3] Let us give an "elementary" proof. We suppose that such an isomorphism $A^n \to A^m$ exists. It is then defined by a matrix $M \in M_{m,n}(A)$. The inverse isomorphism has matrix $N \in M_{n,m}(A)$. These two matrices satisfy

$$MN = \mathrm{Id}_{m,A} \text{ and } NM = \mathrm{Id}_{n,A}.$$

Let $\mathfrak{m}$ be a maximal ideal of A (which is non-zero!) and denote by $k = A/\mathfrak{m}$ the residual field (for the existence of $\mathfrak{m}$, see Corollary 11.1.4). Reducing these matrix identities mod $\mathfrak{m}$, we deduce the existence of matrices in k satisfying

$$\overline{MN} = \mathrm{Id}_{m,k} \text{ and } \overline{NM} = \mathrm{Id}_{n,k}.$$

The matrix $\overline{M}$ thus defines an isomorphism of k-vector spaces $k^n \xrightarrow{\sim} k^m$. The theory of dimension then assures $n = m$. This integer n is called the **rank** of the free module A^n.

This property is false if we no longer assume the ring to be commutative.

---

[3] The reader familiar with exterior algebra will find the statement obvious.

3.8   The Chinese Lemma

We know that the rings $\mathbf{Z}/nm\mathbf{Z}$ and $\mathbf{Z}/n\mathbf{Z} \times \mathbf{Z}/m\mathbf{Z}$ are isomorphic if n and m are coprime. This last condition can also be written as $(n) + (m) = \mathbf{Z}$ according to Bézout's identity.

More generally, suppose we have ideals $I_1, \ldots, I_n$ of a ring A such that $I_i + I_j = A$ for $i \neq j$.

Lemma 3.8.1 (Chinese Lemma)   *Under these conditions, the canonical map* $A \to \prod_{1 \leq j \leq n} A/I_j$ *factors through* $\bigcap_{1 \leq j \leq n} I_j$ *to give an isomorphism*

$$A/(I_1 \cap \cdots \cap I_n) \overset{\sim}{\to} \prod_{1 \leq j \leq n} A/I_j.$$

Moreover, we have

$$I_1 \cap \cdots \cap I_n = I_1 \cdots I_n.$$

Remark 3.8.2 The Chinese Lemma (for integers), known as the Chinese Remainder Theorem, is attributed to the Chinese mathematician and astronomer Sunzi (pinyin spelling) (or Sun Tzu).[4] It seems that his mathematical treatise was written around the year 400 AD (at least that is what the recognized historian of science Qian Baocong wrote in 1963), even if some think he lived around 300 AD. What is certain is that the first written version can be found in the book of Qin Jiushao,[5] *Mathematical Treatise in 9 Sections*, dated 1247.

Proof Of course, the kernel of

$$A \to A/I_1 \times \cdots \times A/I_n$$

is the intersection $I_1 \cap \cdots \cap I_n$. By the universal property of the quotient, we therefore have a map

$$A/I_1 \cap \cdots \cap I_n \to \prod_{1 \leq j \leq n} A/I_j$$

---

[4] Contrary to what one can sometimes read on the web, he has nothing to do with the author of *The Art of War*.

[5] 1202–1261.

which is injective (we have "killed" the kernel of the initial arrow!). Let us verify the surjectivity. If we denote by $I(-j)$ the ideal

$$I(-j) = I_1 \cdots \widehat{I_j} \cdots I_n = \prod_{1 \le k \le n, k \ne j} I_k,$$

that is, the product of ideals I_i distinct from I_j (i.e., generated by the products of elements of the I_i distinct from I_j), we observe that we have

$$\sum_{1 \le j \le n} I(-j) = A.$$

Indeed, we can do an induction on n. If $n = 2$, we have the hypothesis $I_2 + I_1 = A$. Otherwise, we apply the induction hypothesis to $I_1, \ldots, I_{n-1}$. We then obtain that the sum of the $n - 1$ ideals $I_1 \cdots \widehat{I_j} \cdots I_{n-1}$ is A, so that, multiplying by I_n, we have

$$\sum_{1 \le j < n} I(-j) = I_n.$$

and the sum $\sum_j I(-j)$ contains I_n. By applying the same process to $I_2, \ldots, I_n$, we find that the sum contains I_1. As $I_1 + I_n = A$, the sum equals A.

We then write $1 = \sum_{1 \le j \le n} a_j$ with $a_j \in I(-j)$. Let $\bar{b}_j \in A/I_j$ be any classes. We set

$$b = \sum_{1 \le j \le n} a_j b_j.$$

We then observe

$$a_j \equiv \begin{cases} 0 \bmod I_i \text{ if } i \ne j, \\ 1 \bmod I_i \text{ if } i = j, \end{cases}$$

so that $b \equiv b_j a_j \equiv b_j \bmod I_j$ for all j.

It remains to convince ourselves that the product of the I_i, clearly in the intersection of the I_i, is equal to it. Let therefore a be in this intersection. We have $a = \sum_i a a_i$. As $a \in I_i$, we have $a \in I_i I(-i) = I_1 \cdots I_n$ for all i, which is what we wanted. $\square$

Exercise 3.8.3 Let d be a divisor of $n > 0$. Show that the canonical ring morphism

$$\mathbf{Z}/n\mathbf{Z} \to \mathbf{Z}/d\mathbf{Z}$$

induces a surjection at the level of the groups of invertibles (use the Chinese Remainder Theorem and Exercise 3.6.11).

3.9 The Frobenius Morphism

Let p be a prime number and A a ring of characteristic p. We show that A always has a non-trivial endomorphism. This cannot be tautological and has important consequences. Indeed, the ring $\mathbf{R}$ for example does not have a non-trivial endomorphism:

Exercise 3.9.1 Let f be a ring endomorphism of $\mathbf{R}$. Show that the restriction of f to $\mathbf{Q}$ is the identity. Show that f preserves $\mathbf{R}^+$ (study the image of a square). Deduce that f is increasing and then that f is the identity.

More precisely, we prove the following easy and important theorem.

Theorem 3.9.2 (Frobenius Morphism) *The map* $\mathrm{F} : a \mapsto a^p$ *defines an endomorphism of the ring* A.

Proof Clearly, F respects the product and $\mathrm{F}(1) = 1$. We show that F respects the sum. According to the binomial theorem, we have

$$\mathrm{F}(a+b) = \sum_{n=0}^{p} \binom{p}{n} a^n b^{p-n} = \mathrm{F}(a) + \sum_{n=1}^{p-1} \binom{p}{n} a^n b^{p-n} + \mathrm{F}(b).$$

As A is of characteristic p, if $m \in \mathbf{Z}$ is a multiple of p, we have $m\mathrm{A} = 0$. It is therefore sufficient to prove the well-known lemma below.   □

The endomorphism F is called the *Frobenius morphism*, after the mathematician (Fig. 3.1).

Lemma 3.9.3 *Let n be such that $0 < n < p$. Then, the binomial coefficient $\binom{p}{n}$ is divisible by p.*

Proof We have

$$n!\binom{p}{n} = p(p-1)\cdots(p-n+1)$$

(there are n factors). Therefore, p divides the product $n!\binom{p}{n}$. As $n! = n(n-1)\cdots 1$ is a product of integers distinct from the prime p (because $n < p$), it is prime to p. Therefore, the prime number p necessarily divides $\binom{p}{n}$ (this is Gauss's Lemma for integers).   □

Fig. 3.1 Georg Ferdinand
Frobenius (1848–1917).
Photographer: Günther, Carl.
Source: ETH-Bibliothek
Zürich, Bildarchiv, Portr
12004 (ETH, http://doi.org/
10.3932/ethz-a-000046508)

Chapter 4
Basic Concepts of Algebras Over a Field

Algebras and field extensions play a crucial role in Galois theory. In this chapter, we study the definitions and general properties of these structures.

4.1 Algebras and Algebra Morphisms

It is well known that $\mathbf{C}$ is both a field and an $\mathbf{R}$-vector space. Moreover, the structure of external multiplication by the reals is compatible with the product structure of $\mathbf{C}$ in the sense that $x \cdot z$ (external multiplication of the complex number z by the real x) is also the product of the complex numbers $x \cdot 1$ and z (and similarly for the sum). We say that $\mathbf{C}$ is an $\mathbf{R}$-algebra. More generally, if k is a subfield of the field K, then K is naturally a k-vector space and the k-vector space structure on K is compatible with the field structure on K. More generally, we give the following definition.

Definition 4.1.1 Let k be a field and B a ring. We say that B is a *k-algebra* if B is also equipped with an external multiplication $k \times \mathrm{B} \to \mathrm{B}$ making it a k-vector space such that

$$1 \cdot b = b \text{ and } a \cdot (bb') = (a \cdot b)b' \text{ for all } a \in k, \ b, b' \in \mathrm{B}.$$

This last condition is the compatibility condition between the ring and vector space structures. It is equivalent to giving a ring morphism $f : k \to \mathrm{B}$ because we then define the vector space structure by $a \cdot b = f(a)b$ for $a \in k, \ b \in \mathrm{B}$.

Example 4.1.2 The ring of polynomials $k[\mathrm{X}]$ with coefficients in k has a natural structure of a k-algebra.

© The Author(s), under exclusive license to Springer Nature Switzerland AG 2024   41
D. Hernandez, Y. Laszlo, *Introduction to Galois Theory*, Springer Undergraduate
Mathematics Series, https://doi.org/10.1007/978-3-031-66182-2_4

Definition 4.1.3 Let B be an algebra over a field k. If the ring B is a field, we say that B is an *extension* of k.

This extension is then denoted B/k. This definition is compatible with the definition from Sect. 3.3, as discussed there.

Definition 4.1.4 A *morphism* $f : B \to B'$ of k-algebras B, B' is a ring morphism that is also k-linear. We denote the set of algebra morphisms by $\mathrm{Hom}_k(B, B')$.

Two extensions K/k and L/k of k are said to be isomorphic (or k-isomorphic) if there exists a k-algebra morphism $f : K \to L$ that is a k-linear isomorphism. We then write $K \simeq L$ and say that f is an isomorphism of k-algebras.

When a k-algebra morphism $f : A \to B$ is injective, we say that it is a (k-) embedding. We then write $f : A \hookrightarrow B$.

Example 4.1.5 If B is a k-algebra, giving an algebra morphism f from $k[X]$ to B is equivalent to giving the image $b \in B$ of X. Indeed, we will then have

$$f\left(\sum a_i X^i\right) = \sum a_i b^i$$

where $a_i \in k$ and conversely such a formula defines an algebra morphism. Thus,

$$\mathrm{Hom}_k(k[X], B)$$

is canonically identified with B.

More generally, we prove in the same way the following result;

Proposition 4.1.6 *Let $b_1, \ldots, b_n$ be elements of a k-algebra B. There exists a unique morphism of k-algebra $\phi : k[X_1, \ldots, X_n] \to B$ such that for each i, $\phi(X_i) = b_i$.*

If ϕ is surjective, we say that B is *generated* by the b_i and we write $B = k[b_1, \ldots, b_n]$.

Note that if A is a k-algebra and I is an ideal of A, the quotient ring A/I is also a k-vector space (because A and I are k-vector spaces). It is easily verified that we thus obtain a k-algebra structure on A/I. The results for quotient rings adapt word for word to the case of k-algebras. This is particularly the case for the Chinese Lemma 3.8.1.

Example 4.1.7 The quotient $K = \mathbf{R}[X]/(X^2 + 1)$ is isomorphic as an $\mathbf{R}$-algebra to $\mathbf{C}$. Let $\overline{X}$ be the class of X in the quotient K. We have two isomorphisms of $\mathbf{R}$-algebras $\sigma, \overline{\sigma} : K \rightarrow \mathbf{C}$ characterized by $\sigma(\overline{X}) = I$ and $\overline{\sigma}(\overline{X}) = -I$.

Exercise 4.1.8 Describe an isomorphism of $\mathbf{R}$-algebras between $\mathbf{R}[X]/(X^2+X+1)$ and $\mathbf{C}$ on the one hand and between $\mathbf{R}[X]/(X(X+1))$ and $\mathbf{R}^2$ on the other (use the Chinese Lemma 3.8.1).

4.2 The Degree of an Algebra

Definition 4.2.1 Let A be a k-algebra. Its dimension is denoted $[A : k]$ (finite or not) and is called the *degree* of A over k.

The degree is therefore a positive integer or $+\infty$. In the case where A is a field K, we speak of the degree of the extension K/k.

Example 4.2.2 The degree of $\mathbf{C}$ over $\mathbf{R}$ is 2, while the degree of $\mathbf{R}$ over $\mathbf{Q}$ is $+\infty$ (exercise).

For k a field and $P \in k[X]$, we have the formula

$$[k[X]/(P) : k] = \deg(P).$$

Indeed, the images of 1, X, ..., $X^{\deg(P)-1}$ in $k[X]/(P)$ form a basis.

Definition 4.2.3 An extension is said to be *finite* if it is of finite degree.

The next theorem gives an important relation between the degrees of extensions.

Theorem 4.2.4 (Telescopic Base) *Let L be a K-algebra where K is a field containing k so that we have inclusions $k \subset K \subset L$. Let $(\lambda_i)_{i \in I}$ and $(\kappa_j)_{j \in J}$ be bases of L/K and K/k respectively. Then, $(\lambda_i \kappa_j)_{i,j \in I \times J}$ is a base of L/k. In particular, we have*

$$[L : k] = [L : K][K : k].$$

Proof If we have

$$\sum_{i\in I, j\in J} a_{i,j}\lambda_i\kappa_j = \sum_{i\in I}(\sum_{j\in J} a_{i,j}\kappa_j)\lambda_i = 0$$

with $a_{i,j} \in k$ we have $\sum_{j\in J} a_{i,j}\kappa_j = 0$ for all $i \in I$ (freedom of the λ_i over K) and therefore $a_{i,j} = 0$ (freedom of κ_j over k). Moreover, every $l \in L$ can be written as

$$\sum_{i\in I} b_i\lambda_i \text{ with } b_i \in K$$

(λ_i a generator over K) and each b_i can be written as

$$\sum_{j\in J} a_{i,j}\kappa_j \text{ with } a_{i,j} \in k$$

(κ_j a generator over k) so that

$$l = \sum_{i\in I, j\in J} a_{i,j}\lambda_i\kappa_j.$$

$\square$

4.3   Rupture Fields

Let P be a polynomial of $k[X]$, which we assume to be *irreducible* (in the sense of rings defined previously, which corresponds to the usual notion of irreducible polynomial). As $k[X]$ is principal, (P) is maximal (Proposition 3.6.6) and the quotient k-algebra $K = k[X]/(P)$ is a field.

The polynomial P can be considered as having coefficients in K. By construction, $P(\overline{X})$ is the class of P in $K[X]/(P)$, and therefore is null, so that $\overline{X}$ is a root of P in K.

Definition 4.3.1 We say that $K = k[X]/(P)$ is the *rupture field* of P.

Example 4.3.2 The rupture field of $X^2 + 1$ over **R** is isomorphic to **C**.

If x denotes the image of X in K, we obviously have $K = k[x]$.
We have therefore constructed a field extension K/k generated by a root $x \in K$ of P.
This is "the smallest" one in the following sense:

Proposition 4.3.3 *Let* L *be an extension of* k *in which* $P \in k[X]$ *has a root* ξ. *Then the rupture field* K *of* P *embeds in* L *(as a* k-*algebra). Moreover, if* L *is generated by* ξ, *the extensions* K/k *and* L/k *are isomorphic.*

Proof Let $\phi : k[X] \to$ L be the morphism of k-algebra such that $\phi(X) = \xi$. Since $P(\xi) = 0$, the ideal (P) is contained in the kernel $Ker(\phi)$ and therefore ϕ induces a morphism $\overline{\phi} : K \to L$. Since K is a field, this morphism is injective. If in addition L is generated by ξ, the morphism $\overline{\phi}$ is also surjective, so it is an isomorphism. $\square$

The following lemma is easy but fundamental.

Lemma 4.3.4 *Let* L *be an extension of* k *and* $K = k[x]$ *the rupture field of an irreducible* $P \in k[X]$. *Then, the map from* $Hom_k(K, L)$ *to* L *that associates* ϕ *with* $\phi(x)$ *defines a bijection from* $Hom_k(K, L)$ *to the roots of* P *in* L.

In other words, we can identify $Hom_k(K, L)$ and the set of roots of P in L.

Proof To give $\sigma \in Hom_k(k[x], L) = Hom_k(k[X]/(P), L)$ is equivalent to giving $\sigma \in Hom_k(k[X], L)$ which annuls P, by the universal property of the quotient (Remark 3.4.5). In other words, it is equivalent to giving $y = \sigma(X)$ such that $\sigma(P(X)) = P(y) = 0$, according to Example 4.1.5. $\square$

4.4 Algebraic and Transcendental Elements

Let k be a subfield of a field K, that is, an extension K/k.

Definition 4.4.1 An element $x \in K$ is said to be *algebraic* over k if there exists a non-zero $P \in k[X]$ that nullifies x. Otherwise, it is said to be *transcendental* (over k). An extension K/k is said to be *algebraic* if all the elements of K are algebraic (over k).

In the case $k = \mathbf{Q}$, we have the following result.

Proposition 4.4.2 *The set of complex numbers that are algebraic over* $\mathbf{Q}$ *is countable.*

Proof The field **Q** being countable, the set of polynomials of **Q**[X] of fixed degree $n \geq 0$ is countable. The set X_n of all their complex roots is therefore countable, since each of these polynomials has at most n roots. The set of roots $\bigcup_{n\geq 0} X_n$ of all non-zero polynomials of **Q**[X] is therefore countable as a countable union of countable sets. This union is exactly the set of elements of K that are algebraic over **Q**.   □

For example, the real number $\ell = \sum_{n\geq 0} 10^{-n!}$ is transcendental over **Q**: it is the first explicit example of a transcendental number (due to Liouville in 1844). The irrationality of ℓ immediately follows from the elementary properties of periodicity of the decimal expansion of a rational number. The transcendence of ℓ comes as always from the key observation that a certain non-zero positive integer is greater than 1, as illustrated by the following exercise.

Exercise 4.4.3 Let P be a polynomial with integer coefficients P without rational root, d its degree and $x \in \mathbf{R}$ a real root of P. Let $(p, q) \in \mathbf{Z} \times \mathbf{N}^*$.

(1) Show that $d > 1$.
(2) Show that $|P(\frac{p}{q})| \geq \frac{1}{q^d}$.
(3) Show that there exists a C > 0 such that if $\frac{p}{q} \in [x - 1, x + 1]$ then

$$\left| x - \frac{p}{q} \right| \geq \frac{C}{q^d}.$$

(4) Show that ℓ is transcendental over **Q**.

It is well known that the numbers e (Hermite, 1872, Fig. 4.1) and π (Lindemann, 1882, Fig. 4.2) are transcendental over **Q**. A proof of the transcendence of e and π, based on a simplification of the original proofs due to Hilbert largely inspired by [Cha05], will be given later in Sect. 11.3.

Recall that the transcendence of π ensures the unsolvability of a problem that is over 3 millennia old, the squaring of the circle, because otherwise $\sqrt{\pi}$ and therefore π would be algebraic over **Q**.

Fig. 4.1 Charles Hermite
(1822–1901). Author: Cliché
Pirou, Héliog Dujardin, Imp
Ch. Whittmann. Source:
Correspondance d'Hermite et
de Stieltjes, Paris,
Gauthier-Villars, 1905

Fig. 4.2 Ferdinand von Lindemann (1852–1935). Author unknown. Source: Wikimedia Commons (Source St Andrews, https://mathshistory.st-andrews.ac.uk/Biographies/Lindemann/pictdisplay/)

4.5 The Degree of Transcendence

Let K/k be an extension of a field K. We define the algebraic analogue of the notion of a linearly independent family in a vector space.

> **Definition 4.5.1** A family $\mathcal{F} = (x_i)_{i\in I}$ of elements of K is said to be *algebraically independent* (over k) if for all $N \geq 0$, any finite sub-family $\{x_{i_1}, \ldots, x_{i_N}\} \subset \mathcal{F}$ (the i_j are distinct) and any polynomial of N variables $P(X_1, \ldots, X_N) \in k[X_1, \ldots, X_n]$ not equal to zero, we have $P(x_1, \ldots, x_N) \neq 0$.

In other words, there is no non-trivial algebraic relation between the x_i. By analogy with the notion of dimension for a vector space, we define the notion of transcendence degree as follows.

> **Definition 4.5.2** Suppose there exists an $N \geq 0$ such that there is no algebraically free family of K of cardinality $N + 1$ and there exists an algebraically free family of K of cardinality N. We then say that K/k is of *transcendence degree* N.
>   Otherwise, we say that the transcendence degree of K/k is infinite.

For example, the field of fractions with n-indeterminates

$$\text{Frac}(k[X_1, \ldots, X_n]) = k(X_1, \ldots, X_n)$$

is a field of transcendence degree n over k. This is far from immediate (cf. corollary 2 of [Bou23, V.14.3]).

Proposition 4.5.3 *The extension* K/k *is of transcendence degree* 0 *if and only if* K/k *is algebraic.*

Proof If the extension is of transcendence degree 0, then any family with one element $\{x\}$ is not algebraically independent. This means that we have a non-zero polynomial $P(X) \in k[X]$ such that $P(x) = 0$, i.e., that x is algebraic over k. Conversely, if K/k is algebraic, for the same reason, no non-empty family is algebraically independent.   □

The field of real numbers has infinite transcendence degree over the field of rationals (exercise).

4.6   Algebraicity Criteria

The following characterization is as elementary as it is fundamental.

Proposition 4.6.1 *The following statements are equivalent.*

  (i) *x is algebraic over k;*
 (ii) *the algebra k[x] is of finite dimension over k;*
(iii) *the algebra k[x] generated by x is a field.*

Proof

(i)$\Rightarrow$(ii):   if x is algebraic over k, it is the root of a polynomial of degree $d > 0$ and $1, \ldots, x^{d-1}$ generate $k[x]$.

(ii)$\Rightarrow$(iii):   the implication comes from the fact that a domain A which is an algebra of finite dimension over a field K is a field. Indeed, let a be non-zero in such an algebra A. We define $\phi : A \to A$ by $\phi(x) = ax$. Then ϕ is k-linear and, as A is a domain, ϕ is injective. Therefore, ϕ is an isomorphism and $1 \in \text{Im}(\phi)$. The result follows.

(iii)$\Rightarrow$(i):   if $k[x]$ is a field, either x is null, and $x = 0$ is certainly algebraic, or $x^{-1} = P(x) \in k[x]$ and the equation

$$xP(x) - 1 = 0$$

is a linear relation between the x^i, $i \leq \deg(P) + 1$, proving that $k[x]$ is of finite dimension over k.

□

A non-zero polynomial $P \in k[X]$ is said to be *monic* if its leading coefficient is equal to 1.

Proposition 4.6.2 *Let* I *be a non-zero ideal of* $k[X]$. *Then there exists a unique monic polynomial* $P \in k[X]$ *such that* $I = (P)$.

Proof We have already seen the existence of $P \in k[X]$ such that $I = (P)$. If another polynomial Q generates I, then P divides Q and Q divides P, so there exists a non-zero $\lambda \in k$ such that $Q = \lambda P$. Conversely, such a polynomial λP generates I. Therefore, the set of polynomials that generate I is exactly the set of λP with $\lambda \in k$ non-zero. In these sets, there is only one monic polynomial, obtained for λ equal to the inverse of the leading coefficient of P.   □

Definition 4.6.3 Let x be algebraic over k.

We call the *minimal polynomial* of x the monic generator of the ideal of polynomials in $k[X]$ that annihilate x (the ideal of annihilating polynomials of x).

We call the *degree* $\deg_k(x)$ of x over k the dimension $[k[x] : k]$.

Proposition 4.6.4 *Let* P *be the minimal polynomial of* $x \in K$ *algebraic over* k. *Then,*

- P *is irreducible;*
- *the field* $k[x]$ *is (canonically)* k-*isomorphic to* $k[X]/(P)$;
- *we have* $\deg_k(x) = \deg(P)$.

Proof By definition, the algebra morphism $k[X] \to K$ that sends X to x (Example 4.1.5) has $k[x]$ as its image and the ideal (P) as its kernel. We therefore have (Remark 3.4.5) an isomorphism of k-algebras $k[X]/(P) \xrightarrow{\sim} k[x]$ from which the formula $\deg_k(x) = \deg(P)$ follows since the images of the monomials $X^n, 0 \leq n < \deg(P)$ form a basis of $k[X]/(P)$. If now $P = QR$ with Q, R monic, we have $Q(x)R(x) = 0$. As K is a domain, we have $Q(x) = 0$ and therefore $P|Q$. As $\deg(Q) \leq \deg(P)$, we have $P = Q$ and P is irreducible (we can also invoke Proposition 3.6.6 if we want).   □

Definition 4.6.5 Let x be algebraic over k with minimal polynomial P and L an extension of k. The roots of P in L are called the k-*conjugates* of x in L (or *conjugates* in L when the base field k is clear in the context).

Let L be an extension of k and x algebraic over k. Then, $\mathrm{Hom}_k(k[x], \mathrm{L})$ identifies with the conjugates of x in L. More precisely, taking into account Proposition 4.6.4 and Lemma 4.3.4, we obtain the following result.

Proposition 4.6.6 *The map that associates $\sigma(x)$ to $\sigma \in \mathrm{Hom}_k(k[x], \mathrm{L})$ is a bijection between the set of k-embeddings of $k[x]$ into L and the conjugates of x in L.*

Proposition 4.6.7 *The set A of elements of K which are algebraic over k is a subfield of K.*

Proof A and $\mathrm{A} - 0$ are non-empty. Let us verify that the difference and the product of two algebraic x, y is algebraic. By hypothesis, the x^i ($0 \le i \le \deg_k(x)$) and y^j ($0 \le \deg_k(y)$) generate $k[x]$ and $k[y]$ respectively. We deduce that the monomials $x^i y^j$ with $0 \le i \le \deg_k(x), 0 \le j \le \deg_k(y)$ generate $k[x, y]$, which is therefore of finite dimension over k. But $k[x - y]$ and $k[xy]$ are contained in $k[x, y] = k[x][y]$, so they themselves are of finite dimension. If x is non-zero and algebraic, annihilated by P, then $1/x$ is nullified by $\mathrm{X}^{\deg(\mathrm{P})}\mathrm{P}(1/\mathrm{X})$, which is a non-zero polynomial. □

Of course, the degree of a field extension is greater than the degree of all its elements.

As a consequence of Theorem 4.2.4, we have the following result.

Corollary 4.6.8 *If $x_1, \ldots, x_n$ are algebraic, then the algebra $k[x_1, \ldots, x_n]$ of polynomials in the x_i is a field of dimension over k less than or equal to $\prod_{1 \le i \le n} \deg_k(x_i)$.*

We deduce the following result.

Corollary 4.6.9 *An extension of a field k is finite if and only if it is algebraic and generated by a finite number of elements.*

Proof An algebraic extension generated by a finite number of elements is finite according to the previous result. The converse is proved by induction on the degree $[\mathrm{K} : k]$ of such an extension K/k. For $[\mathrm{K} : k] = 1$ it is clear. In general, consider

Fig. 4.3 David Hilbert
(1862–1943). Author
unknown. Source
ETH-Bibliothek Zürich,
Bildarchiv, Dia 326-326

x in K that is not in k. Then $[K : k[x]] < [K : k]$. It is then sufficient to apply the induction hypothesis to the extension $K/k[x]$. We then obtain $x_1, \ldots, x_N$ such that

$$K = (k[x])[x_1, \ldots, x_N] = k[x, x_1, \ldots, x_N].$$

$\square$

Note that the term "generated" in the corollary can mean "generated as a k-algebra" or "generated as a vector space over k" (the statement is true in both cases).

Remark 4.6.10 Conversely, one can prove, but it is more difficult and much deeper, that $k[x_1, \ldots, x_n]$ is a field if and only if it is of finite dimension over k. This is the famous Hilbert's (Fig. 4.3) Nullstellensatz.

4.7   The Concept of Algebraic Closure

A polynomial $P \in K[X]$ is said to be *split* if all its roots are in K, that is, if P can be factored in the form

$$P(X) = \lambda(X - x_1) \cdots (X - x_n)$$

with $\lambda, x_1, \ldots, x_n \in K$.

Fig. 4.4 Joseph Liouville
(1809–1882). Author
unknown. Source: Wikimedia
Commons (Source
Wikipedia, https://upload.
wikimedia.org/wikipedia/
commons/thumb/8/8b/
Joseph_Liouville.jpg/520px-
Joseph_Liouville.jpg)

Definition 4.7.1 We say that a field K is *algebraically closed* if every non-constant polynomial from K[X] is split over K.

The fundamental example is the field **C**, which is algebraically closed. This can be obtained as a consequence of Liouville's (Fig. 4.4) theorem (see [Rud87]):

Theorem 4.7.2 (Liouville) *Let $f : \mathbf{C} \to \mathbf{C}$ be a holomorphic function. If f is bounded then f is constant.*

We deduce:

Corollary 4.7.3 *The field* **C** *is algebraically closed.*

Proof Let $P \in \mathbf{C}[X]$ be a non-constant monic polynomial of degree $n > 0$. Suppose $P(z)$ is non-zero for all $z \in \mathbf{C}$. Then, $1/P$ is holomorphic as a quotient of holomorphic functions with a denominator that does not vanish. Let $a \geq 0$ be the maximum of the moduli of the coefficients of P of degree strictly smaller than n. We have for $|z| > 1$ the inequality $|P(z)|/|z|^n \geq 1 - a/|z|$ and therefore $\lim_{|z| \to \infty} |1/P(z)| = 0$. By continuity, we deduce that $1/P$ is bounded on **C**, therefore constant according to Liouville's theorem, a contradiction.   □

Exercise 4.7.4 Let P be an irreducible polynomial of $k[X]$ and let L be a field containing k that contains a root of P. Show that we can find a k-morphism (injective) from the rupture field of P into L. If P is arbitrary and non-constant, show by induction on $\deg(P)$ that there exists an extension L/k such that P is split over L. Generalize to the case of a family $P_1, \ldots, P_n$ of non-constant polynomials.

Definition 4.7.5 We say that a field K is an *algebraic closure* of the subfield k if K is algebraic over k and if every polynomial in $k[X]$ is split over K.

With this definition, if Ω is algebraically closed and contains k, the set of elements of Ω that are algebraic over k is on one hand a field (Proposition 4.6.7) and on the other is an algebraic closure of k.

Before showing that an algebraic closure always exists, let us prove the following "reassuring" lemma.

Lemma 4.7.6 *An algebraic closure $\bar{k}$ of k is algebraically closed.*

Proof Let $P \in \bar{k}[X]$ be non-constant. It suffices to show that it has a root in $\bar{k}$. The field L generated by the coefficients of P are of finite dimension over k, since the coefficients of P are algebraic over k. Thus, the k-algebra $A = L[X]/(P)$ is of finite dimension over k, namely $\deg(P) \dim_k(L)$ (telescopic base). Consider the morphism $k[T] \to A$ of evaluation in the class x of X in A (that is, which associates P with $P(x)$). This morphism is therefore not injective since $\dim_k(k[T]) = \infty$. Let $Q \in k[T] - 0$ be such that it annihilates x, in other words $P|Q$ and in particular Q is not constant (since $\deg(P) > 0$). But as Q has coefficients in k, Q is split over $\bar{k}$. The same is therefore true of P, which divides it.   □

Example 4.7.7 The subfield $\overline{\mathbf{Q}}$ of $\mathbf{C}$ comprising the elements of $\mathbf{C}$ algebraic over $\mathbf{Q}$ is therefore algebraically closed. But $\overline{\mathbf{Q}}$ is not equal to $\mathbf{C}$! Indeed, we have seen that $\overline{\mathbf{Q}}$ is countable, while $\mathbf{C}$ is not.

Theorem 4.7.8 (Steinitz) *Every field k admits an algebraic closure, unique up to k-isomorphism.*

Fig. 4.5 Ernst Steinitz
(1871–1928). Author
unknown. Source: Archives
of P. Roquette, Heidelberg,
and the Mathematisches
Forschungsinstitut
Oberwolfach

Note that the isomorphism whose existence is asserted by the previous theorem is far from being unique, as we will see: we can even prove that an algebraically closed field admits an infinity of automorphisms. We will first prove the existence of the algebraic closure, then its uniqueness, which follows from the fundamental theorem of morphism extension. We invite the reader to skip this proof of existence at first reading (Fig. 4.5).

4.8 Proof of the Existence of the Algebraic Closure

Once again, we will make use of quotients! Let us first build a gigantic algebra in which every polynomial has a root. We denote by $c(P)$ the leading coefficient of a non-zero polynomial P. The simplest approach is to consider the algebra of polynomials with many indeterminates

$$A = k[X_{P,i}]_{P \in k[X] - 0, 1 \leq i \leq \deg(P)}.$$

For $P \in k[X]$ non-zero and $0 \leq i \leq \deg(P)$, we then denote by $\gamma(i, P) \in A$ the coefficients of the polynomial in X

$$P(X) - c(P) \prod_{i=1}^{\deg(P)} (X - X_{P,i}) \in A[X]$$

recalling that an empty product is equal to 1 so that if P is constant (non-zero), we have $\gamma(0, P) = 0$. Let I be the ideal of A generated by the $\gamma(i, P)$ where P ranges over $k[X] - \{0\}$ and $0 \leq i \leq \deg(P)$. The gigantic algebra is A/I. But it has no reason to be a field: it is not even obvious that it is non-zero. Fortunately, this is the case: we have $I \neq A$.

Otherwise, we would be able to write

$$\sum_{j,P} Q_{P,i_j} \gamma(i_j, P) = 1 \text{ with } Q_{P,i_j} \in A.$$

As the coefficients $\gamma(0, P)$ of the constant polynomials are zero, only polynomials of strictly positive degree contribute to this sum. Choose a field extension K/k such that these non-constant polynomials P (which are infinite number) are split with roots $(x_{P,i})_{1 \le i \le \deg(P)}$ in this extension K (Exercise 4.7.4). Let $\phi : A \to K$ be the morphism of k-algebras sending the corresponding $X_{P,i}$ to $x_{P,i}$ and the other indeterminates to 0. Then ϕ induces a morphism $A[X] \to K[X]$ which sends the corresponding polynomials

$$P(X) - c(P) \prod_{i=1}^{\deg(P)} (X - X_{P,i})$$

to

$$P(X) - c(P) \prod_{i=1}^{\deg(P)} (X - x_{P,i}) = 0$$

by construction, so that

$$\phi(\gamma(i, P)) = 0, \forall i.$$

We deduce that $0 = 1$ in K, which cannot be true since a field is non-zero. The ideal J is therefore proper.

Let J be a maximal ideal of A containing I and L the field A/J (for the existence of J, see Corollary 11.1.4). By construction, every non-constant polynomial $P \in k[X]$ is split over L, its roots being the images of $X_{P,i}$ in A/J. All these roots are algebraic over k. But they generate L as a k-algebra. So we indeed have that L algebraic over k.

4.9 Proof of the Uniqueness of the Algebraic Closure

For uniqueness, let us prove the following statement.

Theorem 4.9.1 (Extension of Morphisms) *Let K, Ω be two extensions of k and suppose K is algebraic over k and Ω is algebraically closed. Then, there exists an embedding (of k-algebras) $K \hookrightarrow \Omega$.*

Proof Let E be the (non-empty) set of pairs (L, σ) where L is a subfield of K containing k and σ is a k-embedding

$$\sigma : L \hookrightarrow \Omega.$$

Each σ makes Ω an L-algebra. The extension of such embeddings defines an order relation on E which clearly makes it an inductive set (see Sect. 11.1). Let (L, σ) be a maximal element (Zorn's Lemma, again see Sect. 11.1). Let us show $L = K$. Let $x \in K$. As x is algebraic over k it is also algebraic over L. Let $P(X) = \sum a_i X^i$ be the minimal polynomial of x over L. There is a natural isomorphism of L-algebras between $L[X]/(P)$ and $L[x]$. Let y be a root of $P^\sigma(X) = \sum \sigma(a_i) X^i$ in Ω. There exists a unique morphism of L-algebras $L[X]/P \to \Omega$ such that the image of X is y. Indeed, the image of P in Ω is by definition $P^\sigma(y) = 0$ (Proposition 4.6.6). We therefore now obtain the existence of a morphism of k-algebras $L[x] \to \Omega$ extending σ. By maximality of L, we deduce $x \in L$. □

Remark 4.9.2 σ allows us to identify L with $\sigma(L)$. From now on, we will do this directly, without distinguishing between L and $\sigma(L)$ (cf. Sect. 3.3). Also note that if K is a finite extension, then the notion of dimension allows us to show the existence of L without recourse to Zorn's Lemma.

Another way of expressing the statement of the extension theorem, which we will often use, is the following.

Corollary 4.9.3 *Let K/k be algebraic and Ω/k be algebraically closed. Let σ : $K \to \Omega$ be a morphism of k-algebras and Ω'/k be an algebraic closure of k containing K. Then σ can be extended to a morphism of k-algebras $\tilde{\sigma} : \Omega' \to \Omega$.*

Proof The field Ω' is a K-algebra that is algebraic over K. Moreover, σ makes Ω a K-algebra that is algebraically closed over K. Therefore, according to the extension theorem, we have an embedding $\Psi : \Omega' \to \Omega$ that is a morphism of K-algebras. But then as $k \subset K$, Ψ is also a morphism of k-algebras. Furthermore, for $x \in K$, we have $\Psi(x \cdot 1) = \sigma(x)\Psi(1)$, so $\Psi = \sigma$ on K. Thus, Ψ indeed extends σ. □

We now have the following important consequences.

Corollary 4.9.4 *Two algebraic closures K_1, K_2 of k are k-isomorphic.*

Proof Considering K_1 as algebraic and K_2 as algebraically closed, the extension theorem 4.9.1 ensures that there is an embedding of K_1 into K_2. With the previous notations, the choice of such an embedding of K_1 into K_2 allows us to view K_1 as a subfield of K_2. Then replacing the roles of (k, K, Ω) with (K_1, K_2, K_1), we deduce the existence of $\tau \in \mathrm{Hom}_{K_1}(K_2, K_1)$, in other words a commutative diagram

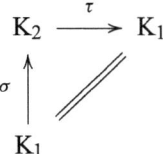

(the term "commutative diagram" means that the two directions of the diagram give the same map, that is $\tau \circ \sigma = \mathrm{Id}$). As τ is a field morphism, τ is injective. The equality $\tau \circ \sigma = \mathrm{Id}$ ensures its surjectivity; τ and σ are inverses of each other.   □

Corollary 4.9.5 *Let* K/k, Ω/k *be two extensions of* k *with* K/k *algebraic and* Ω *algebraically closed. Then, the conjugates in* Ω *of* $x \in K$ *are the* $\sigma(x)$, $\sigma \in \mathrm{Hom}_k(K, \Omega)$.

Proof If $y \in \Omega$ is a conjugate of x, there exists a $\sigma \in \mathrm{Hom}_k(k[x], \Omega)$ such that $\sigma(x) = y$ (Proposition 4.6.6). It remains to extend σ to K as a whole, which is possible (Theorem 4.9.1). Conversely, $\sigma \in \mathrm{Hom}_k(K, \Omega)$ leaves invariant the minimal polynomial of x over k (that is, σ leaves invariant all its coefficients). It therefore permutes its roots, which are the conjugates of x by definition (Proposition 4.6.6).   □

4.10   The Splitting Field of a Polynomial

Let k be a subfield of an algebraically closed field Ω. Let P be a polynomial of $k[X]$, not necessarily irreducible.

Definition 4.10.1 The *splitting field* of P is the subfield of Ω generated by the roots of P in Ω.

This is the smallest subfield of Ω in which P is split. Of course, it is contained in the algebraic closure of k in Ω, a subset of the algebraics over k. From this it follows that it does not depend on Ω, up to non-unique isomorphism (Corollary 4.9.4). This justifies the use of "the" in the expression **the splitting field of** P. In the following, we will thus fix an algebraic closure in which we will work. This allows us to talk about the splitting field. For example, we will be interested in subfields of **C** algebraic over **Q**.

Chapter 5
Finite Fields and Perfect Fields

The framework chosen in this book is that of perfect fields, which we study in this chapter. We first introduce finite fields, which are fundamental examples of perfect fields.

5.1 Existence and Uniqueness of Finite Fields

Let k be a finite field (that is, of finite cardinality as a set). We have seen (Proposition 3.5.4) that the characteristic of k is necessarily a prime number $p > 0$ and that it contains the field $\mathbf{F}_p = \mathbf{Z}/p\mathbf{Z}$. Let us study the cardinality of k.

Lemma 5.1.1 *The cardinality of a finite field k is of the form $q = p^n$ where p is the characteristic of k.*

Proof Since k is finite, it has finite dimension n as a vector space over $\mathbf{F}_p$. The choice of a basis defines an $\mathbf{F}_p$-isomorphism of vector spaces $\mathbf{F}_p^n \xrightarrow{\sim} k$. Since $\mathbf{F}_p^n$ has cardinality p^n, the lemma is proven. $\qquad\square$

Let Ω be an algebraic closure of $\mathbf{F}_p$. Since k has finite cardinality, it is algebraic over $\mathbf{F}_p$. According to Theorem 4.9.1, k embeds into Ω. We can therefore assume that k is contained in Ω.

Lemma 5.1.2 *k is the set of roots in Ω of the polynomial $X^q - X$.*

Proof Since k^* is of order $q - 1$, we have $x^{q-1} = 1$ for all $x \neq 0$ according to Lagrange's theorem. Therefore $x^q = x$ for all $x \in k$. Since the polynomial $X^q - X$ admits at most $\mathrm{card}(k) = q$ roots, we deduce that k is necessarily the set of roots of $X^q - X$. $\qquad\square$

The lemma implies in particular the uniqueness of a field of cardinality $q = p^n$, in the sense that any subfield of Ω of cardinality q is equal to k.

D. Hernandez, Y. Laszlo, *Introduction to Galois Theory*, Springer Undergraduate Mathematics Series, https://doi.org/10.1007/978-3-031-66182-2_5

Lemma 5.1.3 *Two finite fields are isomorphic if and only if they have the same cardinality.*

Proof Let k, k' have the same cardinality $q = p^n$. Choose Ω algebraically closed of characteristic p (thus containing $\mathbf{F}_p$). The identity of $\mathbf{F}_p$ extends to embeddings s, s' of k, k' into Ω. But as Ω has a unique subfield of cardinality q, we have $s(k) = s'(k')$, so that $s^{-1}s'$ is well defined and is the sought isomorphism.   □

We now show the existence.
Consider therefore $\mathbf{F}_q$, the *set* of roots in Ω of $X^q - X$.

Lemma 5.1.4 $\mathbf{F}_q$ *is a subfield of Ω with q elements (and it is the only one).*

Proof Let us denote by F the Frobenius morphism (Theorem 3.9.2) $\mathrm{F} : x \mapsto x^p$ of Ω.

Recall that F is a ring morphism, and therefore so is its iterate F^n. We therefore have

$$(x + y)^q = \mathrm{F}^n(x + y) = \mathrm{F}^n(x) + \mathrm{F}^n(y) = x^q + y^q,$$

which proves the stability of $\mathbf{F}_q$ under sums. The stability under product, inverse and negation is obvious. Thus, $\mathbf{F}_q$ is a subfield. It remains to show it has cardinality q. We need to prove that the roots are simple. If one of them was at least double, it would also be a root of $(X^q - X)'$, but the latter is -1. The uniqueness was seen in the discussion at the beginning of this section: such a field is necessarily the set of roots of $X^q - X$.   □

Lemma 5.1.5 $\mathbf{F}_{p^n}$ *is contained in $\mathbf{F}_{p^m}$ if and only if $n|m$. Moreover, for $q = p^n$, the algebraic closure $\overline{\mathbf{F}}_q$ of $\mathbf{F}_q$ in Ω is the increasing union $\bigcup_{N \geq 1} \mathbf{F}_{q^{N!}}$.*

Proof If $n|m$, any root of $X^{p^n} - X$ is a root of $X^{p^m} - X$, hence the inclusion $\mathbf{F}_{p^n} \subset \mathbf{F}_{p^m}$. Conversely, if $\mathbf{F}_{p^n} \subset \mathbf{F}_{p^m}$, we have $\mathbf{F}^*_{p^n} \subset \mathbf{F}^*_{p^m}$ and therefore $(p^n - 1)|(p^m - 1)$ according to Lagrange's theorem. Let us write the Euclidean division as $m = an + r$, $0 \leq r < n$. We then have

$$p^m - 1 = p^{an}p^r - 1 = (p^{an} - 1)p^r + p^r - 1.$$

But, according to the formula for the partial sums of a geometric series, $(p^n - 1)|(p^{an} - 1)$. Thus $p^n - 1|p^r - 1 < p^n - 1$, which is possible only if $r = 0$. We obtain the desired result.

We now consider $q = p^n$. For $N \geq 1$, the elements of $\mathbf{F}_{q^{N!}}$ are algebraic over $\mathbf{F}_q$ and therefore $\mathbf{F}_{q^{N!}} \subset \overline{\mathbf{F}}_q$.

Let $x \in \overline{\mathbf{F}}_q$. Then we have a non-trivial polynomial $P(X) \in \mathbf{F}_q[X]$ such that $P(x) = 0$. By applying the iterates of F^n (the Frobenius morphism to the power n), we obtain $P(F^{rn}(x)) = 0$ for all $r \geq 0$. Therefore $(F^{rn}(x))_{r \geq 0}$ is included in the set of roots of P, which is finite. Therefore there exists $r' > r \geq 0$ such that $F^{r'n}(x) = F^{rn}(x)$. This implies $F^{(r'-r)n}(x) = x$ and therefore $x \in \mathbf{F}_{q^{r'-r}}$.   □

Note that, for $n|m$, if d is the dimension of $\mathbf{F}_{p^m}$ over $\mathbf{F}_{p^n}$, we have, as a vector space, $\mathbf{F}_{p^m} \xrightarrow{\sim} (\mathbf{F}_{p^n})^d$. By counting the cardinals, we obtain $p^m = (p^n)^d$, hence $d = m/n$.

Also note that, *a fortiori*, $\mathbf{F}_q$ is the splitting field of $X^q - X$ over $\mathbf{F}_p$ (in Ω). We will therefore talk about the finite field $\mathbf{F}_q$ (it is generally implied that an algebraically closed field of characteristic p has been chosen).

5.2 Automorphisms of Finite Fields

We will use the following classic result.

Proposition 5.2.1 *Let k be a field. Every finite subgroup of k^* is cyclic.*

Proof Let $G \subset k^*$ be a finite subgroup of order $n = |G|$. Then, according to Lagrange's (Fig. 5.1) theorem, we have $x^n = 1$ for all $x \in G$. Therefore $X^n - 1$ is split in $k[X]$ with distinct roots, its roots being exactly the n elements of G. Now for d an integer that divides n, $X^d - 1$ divides the polynomial $X^n - 1$. Therefore $X^d - 1$ is split with distinct roots. For example, let p be a prime number that divides n and $r \geq 1$ maximal such that $d = p^r$ divides n. Then $X^d - 1$ has d distinct

Fig. 5.1 Joseph Louis Lagrange (1736–1813). Author unknown. Source: ETH-Bibliothek Zürich, Bildarchiv, Dia 326-409

roots in k. The roots of order different from d are roots of $X^{p^{r-1}} - 1$, so there are at most p^{r-1} of them. As $p^r > p^{r-1}$, there is at least one root $x \in k$ of $X^d - 1$ that is not root of $X^{p^{r-1}} - 1$, and therefore x is of order p^r. Now let $x_1, \ldots, x_N$ be obtained in this way for each prime number that divides n. Let $y = x_1 \cdots x_N$. As G is commutative and for $i \neq j$ the order of x_i is coprime with that of x_j, the order of y is the product of the orders of $x_1, \ldots, x_n$, that is n. Therefore G is cyclic.   □

For example, for p a prime number, $(\mathbf{Z}/p\mathbf{Z})^*$ is a cyclic group of order $p - 1$.

Remark 5.2.2 If we know the structure of finite abelian groups, this result is obvious. Indeed, we then know that k^* is isomorphic to a product

$$\Pi = \prod_{i=1}^{d} \mathbf{Z}/n_i\mathbf{Z}$$

with $1 < n_1 | \cdots | n_d$ (note, the law on k^* is multiplicative, while on the right the law is additive with neutral element 0). However, in a field, the number of solutions to $X^{n_1} = 1$ is at most n_1. In Π, they correspond to the solutions of the equation $n_1\pi = 0$. If $d > 1$, there are at least $2n_1$, namely the elements of $\mathbf{Z}/n_1\mathbf{Z}$ and those of

$$n_2/n_1\mathbf{Z}/n_2\mathbf{Z} \xrightarrow{\sim} \mathbf{Z}/n_1\mathbf{Z},$$

a contradiction.

Let $q = p^n$ be a power of a prime number and m an integer > 0. We denote by

$$F_q : \mathbf{F}_{q^m} \to \mathbf{F}_{q^m}$$

the iteration F^n of the Frobenius morphism: $F_q(x) = x^q$ for $x \in \mathbf{F}_{q^n}$. It is a field morphism, which is the identity on $\mathbf{F}_q$ (the set of roots of $X^q - X$).

Theorem 5.2.3 *The group* $\mathrm{Aut}_{\mathbf{F}_q}(\mathbf{F}_{q^m})$ *is cyclic of order m generated by* F_q.

Proof Let x be a generator of the cyclic group $\mathbf{F}_{q^m}^*$. We have *a fortiori* $\mathbf{F}_{q^m} = \mathbf{F}_q[x]$. As $[\mathbf{F}_{q^m} : \mathbf{F}_q] = [\mathbf{F}_q[x] : \mathbf{F}_q] = m$, the minimal polynomial P of x over $\mathbf{F}_q$ is of degree m. A morphism $\sigma \in \mathrm{G} = \mathrm{Aut}_{\mathbf{F}_q}(\mathbf{F}_{q^m})$ leaves P invariant so that $\sigma(x)$ is a root of P, which has at most m roots in k. As x generates $\mathbf{F}_{q^m}^*$, the morphism σ is determined by $\sigma(x)$ so that $\mathrm{card}(\mathrm{G}) \leq m$. Moreover, F_q is of order m. Otherwise, there would exist $0 < d < m$ such that $F^d = \mathrm{Id}$, and therefore $x^{q^d} = x$ contradicting that x is of order $q^m - 1$. However, F_q is indeed an automorphism, since it is an injective map (like any field morphism) between finite sets of the same cardinality.   □

Exercise 5.2.4 (Difficult) Let M $\in$ GL$_n(\mathbf{F}_q)$, regarded as a bijection of $(\mathbf{F}_q)^n$. What is its signature [consider the case of even or odd q separately]? Show that the minimal polynomial of F$_q$ viewed as an endomorphism of the $\mathbf{F}_q$-vector space $\mathbf{F}_{q^n}$ is $X^n - 1$ [prove that distinct homomorphisms from a group G into the multiplicative group k^* of a field k are linearly independent, regarded as functions from G into k]. What is its signature?

Exercise 5.2.5 Show that for any positive integer d, there exists at least one irreducible polynomial P $\in$ $\mathbf{F}_q[X]$ of degree d. Show that P divides $X^{q^d} - X$ and that its rupture field over $\mathbf{F}_q$ is its splitting field over $\mathbf{F}_q$.

5.3   An Application of the Chinese Lemma: The Berlekamp Algorithm

We will provide an algorithm, which can be run on a computer, to factorize a polynomial P $\in$ $\mathbf{F}_p[X]$ into irreducible factors, or at least to determine whether it is irreducible or not.

We are given a prime p and a non-constant monic P $\in$ $\mathbf{F}_p[X]$. Recall the equality

$$x^p = x, \forall x \in \mathbf{F}_p.$$

5.3.1   Where We Reduce to P Without Square Factor

If P is divisible by a square Q^2 of degree > 0, the degree of S $=$ GCD(P, P$'$) is at least deg(Q) and therefore is strictly positive.

If deg(S) $=$ deg(P), then P$'$ is zero since S|P$'$, in other words P can be written as $\sum_i a_{ip}^p X^{ip} = R$ with R $= (\sum_i a_{ip} X^i)^p \in \mathbf{F}_p[X]$, (cf. the proof of Theorem 5.5.3 *infra*). We apply the algorithm again to R.

Otherwise, S $=$ GCD(P, P$'$) is a non-trivial divisor of P and we apply the algorithm to S and P/S, which are of smaller degree.

We can assume, which we do from now on, that P is without square factor.

5.3.2   Fixed Points of the Frobenius Morphism

Write

$$P = \prod_{i=1}^{m} P_i$$

with P_i monic irreducible and pairwise distinct. As P_i and P_j are coprime for $i \neq j$, the sum of the ideals they generate is all of $\mathbf{F}_p[X]$. The Chinese Lemma (3.8.1) ensures that the canonical morphism of $\mathbf{F}_p$-algebras

$$\gamma \,:\, A = \mathbf{F}_p[X]/(P) \rightarrow \oplus_i \mathbf{F}_p[X]/(P_i)$$

is an isomorphism of algebras.

Let F be the Frobenius morphism of A (we will use the same notation for the Frobenius morphism of $A_i = \mathbf{F}_p[X]/(P_i)$). As $\mathbf{F}_p[X]$ is principal and P_i irreducible, each A_i is a finite field (Proposition 3.6.6) so that we have $A_i^F = \mathrm{Ker}(F-\mathrm{Id}_{A_i}) = \mathbf{F}_p$ (Lemma 5.1.4). The formula

$$\gamma(a^p) = \gamma(a)^p = (a_i^p \bmod P_i)$$

ensures that the image of $A^F = \mathrm{Ker}(F - \mathrm{Id}_A)$ under γ is equal to $\mathbf{F}_p^m = \oplus_i \mathbf{F}_p$. In particular, we have

$$\dim_{\mathbf{F}_p} A^F = m.$$

Thus, P is irreducible if and only if $A^F = \mathbf{F}_p$. Note that the calculation of this dimension is perfectly **algorithmic**: we calculate the matrix of F in the base of the classes of X^i, $i = 0, \ldots, d - 1$ (which is done by dividing X^{ip} by P) then we calculate the rank of $F - \mathrm{Id}$ by Gaussian elimination.

We therefore have an algorithmic criterion to determine if P is irreducible, which can be summarized as follows: P **without square factor is irreducible if and only if the matrix of** $F - \mathrm{Id}$ **is of rank** $\deg(P) - 1$, where the latter is calculated using Gaussian elimination.

5.3.3   Factorization of P

Let us go further, in the case $\dim A^F > 1$. In this case, there exists an $a \in A$ which is not in our vector line $\mathbf{F}_p \subset \mathbf{F}_p[X]$ of constant polynomials. In other words, there exists a Q of degree $0 < \deg(Q) < \deg(P)$ such that $\overline{Q} \in A^F$, i.e. $P|F(Q) - Q = Q^p - Q$. From the factorization

$$X^p - X = \prod_{i \in \mathbf{F}_p} (X - i),$$

we get

$$Q^p - Q = \prod_{i \in \mathbf{F}_p} (Q - i)$$

and therefore

$$P \mid \prod_{i \in \mathbf{F}_p} (Q - i).$$

Note that if $i \neq j$ in $\mathbf{F}_p$, the identity

$$1/(j - i)((Q - i) - (Q - j)) = 1$$

ensures $\mathrm{GCD}(Q - i, Q - j) = 1$.

Thus, each factor P_j of P divides **exactly one** of the factors $Q - i$ so that

$$P = \prod_{i \in \mathbf{F}_p} \mathrm{GCD}((Q - i), P).$$

Now, each polynomial $\mathrm{GCD}((Q-i), P)$ (which is calculated thanks to the Bézout algorithm) is of degree strictly less than $\deg(P)$ by construction and we repeat the process for each polynomial $\mathrm{GCD}((Q-i), P)$. This process stops in a finite number of steps.

This process is algorithmic, but not very efficient. Indeed, imagine for example that P is of degree 1000 and p of the order of 10^6. The probability that $\mathrm{GCD}((Q - i), P)$ is different from 1 is of the order of $1/1000$ and we therefore see that this product with 10^6 terms has very few non-trivial factors. In fact, in practice, we adapt this algorithm, which becomes probabilistic.

An exercise is to program this algorithm with a symbolic computation software. Another exercise is to evaluate the number of necessary operations: indeed, as the number of polynomials of given degree in $\mathbf{F}_p[X]$ is finite, we could have performed all the products of two polynomials and compared with P, which gives a factorization algorithm. But, as soon as the degree is large, the number of operations is enormous and blows up any machine. On the other hand, for small p, d, it is efficient. Anyway, the Berlekamp algorithm, relatively inefficient in general, is theoretically interesting.

Remark 5.3.1 The reader should generalize the algorithm, replacing $\mathbf{F}_p$ by $\mathbf{F}_{p^n}$, simply by replacing F with the composite $F_{p^n} = F^n$.

5.4 Extensions of Perfect Fields

We saw in the proof of Theorem 5.2.3 that the Frobenius morphism of a finite field $k = \mathbf{F}_q$ ($q = p^n$) is surjective, in other words, every element of $\mathbf{F}_q$ is a p-th power. However, this is not the case for all fields of positive characteristic. For example, let $k = \mathrm{Frac}(\mathbf{F}_p[t])$ be the field of fractions of $\mathbf{F}_p[t]$. Then, for clear reasons of degree, the element t is not a p-th power.

Definition 5.4.1 We say that a field k is *perfect* if its characteristic is zero or if it is of characteristic $p > 0$ and its Frobenius morphism is surjective.

Note that the Frobenius morphism is always injective, so the condition in characteristic $p > 0$ is equivalent to saying that the Frobenius morphism is an isomorphism.

Example 5.4.2 Finite fields are perfect.

We have the following fundamental example:

Lemma 5.4.3 *An algebraically closed field is perfect.*

Proof Let Ω be such a field. If its characteristic is zero, the result is clear. Otherwise, its characteristic is $p > 0$. But then for every $x \in \Omega$, the polynomial $X^p - x$ has at least one root in Ω, so x is a p-th root. Therefore, Ω is perfect.   □

It is not true that a subfield of a perfect field is perfect. Indeed, the algebraic closure of a non-perfect field (for example $\mathrm{Frac}(\mathbf{F}_p[t])$) is a perfect field containing a subfield that is not. It is also not true that an extension of a perfect field is perfect. For example, the perfect field $\mathbf{F}_p$ has an extension $\mathrm{Frac}(\mathbf{F}_p[t])$ which is not a perfect field. However, we have the following important statement.

Proposition 5.4.4 *Let K/k be an algebraic extension. If k is perfect, then so is K.*

Proof The question only poses a problem in characteristic $p > 0$.

Suppose first that K/k is finite. Let F be the Frobenius morphism of K, which is bijective on k by hypothesis (k is perfect). We therefore have an inverse F^{-1} : $k \to k$. Then, $F(K)$ is clearly an $F(k)$-subspace of K, hence a k-subspace, since $F(k) = k$. Similarly, if the $x_i \in K$ are free over k, the $F(x_i)$ are free over k in $F(K)$. Indeed, if $\sum a_i F(x_i) = 0$ with the $a_i \in k$, then $F(\sum F^{-1}(a_i)x_i) = 0$ and therefore $\sum F^{-1}(a_i)x_i = 0$. The family of x_i being free, this implies $F^{-1}(a_i) = 0$ for all i, and therefore $a_i = 0$. We deduce, by taking a basis, the inequality $[K : k] \leq [F(K) : k]$, then the reverse inequality because $F(K) \subset K$. Therefore $K = F(K)$.

Suppose K/k is algebraic and let $x \in K$. As $k[x]/k$ is finite, the field $k[x]$ is perfect and therefore x is a p-th power in $k[x]$ and therefore in K.   □

The converse is true in the finite case (although less useful). To prove it, let us start with a lemma. Let k be a subfield of an algebraically closed field Ω of characteristic $p > 0$. As the Frobenius morphism of Ω is bijective (Ω is perfect), the p-th root $x^{1/p} = F^{-1}(x)$ of any element of Ω is well defined. The same is true for the p^n-th roots.

Lemma 5.4.5 *Suppose* $t \in k$ *is not a p-th root in k. Then, for all* $n \geq 1$, *the polynomial* $X^{p^n} - t$ *is irreducible in* $k[X]$.

In other words, we have $\deg_k(x^{1/p^n}) = p^n$.

Proof Let τ be the p^n-th root of t. We know that $\tau \notin k$ since by hypothesis

$$t^{1/p} = \tau^{p^{n-1}} \notin k.$$

Let P be the minimal polynomial of τ over k: it is an irreducible polynomial (Proposition 4.6.4) of $k[X]$ which divides $Q = X^{p^n} - t$ since $Q(\tau) = 0$. Using a decomposition of Q into irreducible factors in $k[X]$, we can write:

$$Q = P^m R \text{ with } R \in k[X] \text{ and } GCD(P, R) = 1.$$

Bézout's identity is written as $PU + RV = 1$ with $U, V \in k[X]$. Consequently, P and R do not have a common root in Ω. However, in $\Omega[X]$, we have $Q(X) = (X - \tau)^{p^n}$ so that R has no root at all and therefore $Q = P^m$. By comparing degrees, we get $m = p^\nu, 0 \leq \nu \leq n$. By evaluating at 0, we then have

$$Q(0) = -t = (P(0))^{p^\nu}.$$

Since t is not a p-th power in k, we therefore have $\nu = 0$. Thus $Q = P$ is irreducible. $\square$

We then obtain the converse of Proposition 5.4.4 in the finite case.

Corollary 5.4.6 *Let* K/k *be a finite extension. If* K *is perfect, then* k *is perfect.*

Proof Indeed, if $t \in k$ is not a p-th power, $t^{p^{-n}} \in K$ is of degree p^n over k, which goes to infinity with n. This is a contradiction, because K/k being a finite extension, the degree of the elements of K over k is bounded by $[K : k]$. $\square$

5.5 Separable Polynomials and Perfect Fields

The interest of perfect fields comes from the fundamental Theorem 5.5.3 below. Let k be a field and Ω an algebraically closed field containing it.

Definition 5.5.1 A monic polynomial is said to be *separable* if its roots in Ω are simple.

We recall the following well-known lemma.

Lemma 5.5.2 *A polynomial is separable if and only if it is coprime with its derivative.*

Proof Suppose P is separable. Write

$$P = a \prod_{i \in I} (X - z_i), a \in k^*, z_i \in \Omega,$$

where the z_i are pairwise distinct. We therefore deduce

$$GCD(P, P') = \prod_{i \in I'} (X - z_i) \text{ where } I' \subset I.$$

Suppose $I' \neq \varnothing$ and choose $i' \in I'$. We therefore have $P(z_{i'}) = 0$. However,

$$P' = a \sum_{i \in I} \prod_{j \neq i} (X - z_j)$$

so that

$$\prod_{j \neq i'} (z_{i'} - z_j) = 0,$$

which is absurd because the z_i are pairwise distinct.

Conversely, if P, P' are coprime, the Bézout identity $AP + BP' = 1$ with A, B polynomials ensures that P, P' do not have common roots in Ω and therefore that the roots of P in Ω are simple.   □

Theorem 5.5.3 *A field k is perfect if and only every irreducible polynomial of k[X] is separable.*

Proof Suppose k perfect and let P be an irreducible polynomial of $k[X]$ (in particular non-constant). Let us show that P is coprime with P'. Since P is irreducible, the GCD of P and P' is 1 or P. We show by contradiction that it is 1. If it is P, for degree reasons, it is because P' is the zero polynomial. This already imposes that the characteristic of k is $p > 0$. By writing

$$P = \sum a_n X^n \text{ and } P' = \sum n a_n X^{n-1}$$

we deduce that $na_n = 0$ for all n and therefore $a_n = 0$ if p does not divide n. Thus, we have

$$P = \sum_n a_{np} X^{np}.$$

Since k is perfect, the Frobenius morphism F is bijective and we therefore have

$$P = \sum_n F^{-1}(a_{np}^p) X^{np} = (\sum_n F^{-1}(a_{np}) X^n)^p$$

since $pk[X] = 0$. This is absurd because P is irreducible.

Conversely, suppose that every irreducible polynomial of $k[X]$ is separable. We can suppose k is of characteristic $p > 0$ and show that every element t has a p-th root. Indeed, let us prove the contrapositive; suppose $t \in k$ does not have a p-th root in k. Let $t^{1/p}$ be a p-th root of t in Ω and P its minimal polynomial over k. Then P is irreducible (Proposition 4.6.4), it divides $X^p - t$ and $\deg(P) > 1$ because by hypothesis $t^{1/p} \notin k$ (note that according to the previous section, $X^p - t$ is irreducible in $k[X]$). In $\Omega[X]$, $X^p - t$ is written as $(X - t^{1/p})^p$ and therefore has only one root (of multiplicity p). Thus $P = (X - t^{1/p})^i$ with $2 \le i \le p$. We deduce that P is not separable, a contradiction.   □

5.6   The Primitive Element Theorem

Let k be a perfect field.

Definition 5.6.1 We say that a field extension K/k is *simple* if it can be generated by a single element, that is, if there exists an $x \in K$ such that $K = k[x]$. Such a generator x of the extension is called a *primitive element*.

We then have the following fundamental theorem.

Theorem 5.6.2 (Primitive Element)   *Every finite extension K/k is simple.*

Proof If k is finite, so is K, and K* is cyclic (Proposition 5.2.1), generated by x let's say. We then have $K = k[x]$. So let us assume k is infinite. By induction, we immediately reduce to proving that if x, y are elements of K, there exists a $z \in K$ such that

$$k[z] = k[x, y].$$

We look for z in the form $z = x+ty, t \in k^*$. Let $L = k[z]$. It suffices to prove $x \in L$, because then $y = (z - x)/t \in L$. Let $P_x, P_y \in k[X]$ be the minimal polynomials of x, y over k. The polynomial

$$Q(X) = P_y((z - X)/t)$$

has coefficients in L and cancels x by construction. Let

$$R = GCD(Q, P_x) \in L[X].$$

As we have already noted, the Euclidean algorithm proves that the GCD of polynomials is invariant under a change of field. Thus, for example, this calculation can be done in Ω. Since P_x has simple roots, we have

$$R(X) = \prod_{\substack{x' \text{ such that} \\ Q(x')=P_x(x')=0}} (X - x').$$

If we write $P_y = \prod(X - y')$, the roots of Q are written

$$z - ty' = x + t(y - y').$$

Let us choose $t \neq 0$ outside of the finite number of t such that there exist $y' \neq y$ and x' satisfying

$$x' = x + t(y - y'), \text{ i.e. } t = \frac{x' - x}{y - y'}.$$

Then the set

$$\{x' | Q(x') = P_x(x') = 0\}$$

is reduced to x.

We deduce that having chosen such a t, we have $R(X) = X - x$. As $R \in L[X]$, we have $x \in L$. □

For example, consider the extension $K = \mathbf{Q}(\sqrt{2}, \sqrt{3})$ of $\mathbf{Q}$. Let us show that $\sqrt{2} + \sqrt{3}$ is primitive while $\sqrt{2}, \sqrt{3}$ are not.

Since $\sqrt{2}, \sqrt{3} \notin \mathbf{Q}$, we have

$$[\mathbf{Q}(\sqrt{2}) : \mathbf{Q}] = [\mathbf{Q}(\sqrt{3}) : \mathbf{Q}] = 2$$

and therefore $[K : \mathbf{Q}] = 2$ or 4. If the degree is 2, then $K = \mathbf{Q}(\sqrt{2})$. Then $\sqrt{3} = a\sqrt{2} + b$ with $a, b \in \mathbf{Q}$, which implies $3 - 2a^2 - b^2 = 2ab\sqrt{2}$. Therefore $a = 0$ or $b = 0$ and since $\sqrt{3} \notin \mathbf{Q}$, $b = 0$. Therefore $\sqrt{3/2} \in \mathbf{Q}$, a contradiction. Therefore $[K : \mathbf{Q}] = 4$ and $\sqrt{2}, \sqrt{3}$ are not primitive. On the other hand $x = \sqrt{2} + \sqrt{3}$ is primitive. Indeed, let $L = \mathbf{Q}[x] \subset K$. Then $x^{-1} = \sqrt{3} - \sqrt{2} \in L$. We deduce $\sqrt{3}, \sqrt{2} \in L$ and therefore $L = K$.

Here is another important example.

Lemma 5.6.3 *For integer $n \geq 1$, let $\zeta_n = \exp(\frac{2\mathrm{I}\pi}{n})$. Then for integers $n, m \geq 1$, we have $\mathbf{Q}(\zeta_n, \zeta_m) = \mathbf{Q}(\zeta_{\mathrm{LCM}(n,m)})$.*

Proof Let $\varpi = \mathrm{LCM}(n, m)$. Since $\zeta_\varpi^{\varpi/n} = \zeta_n$, we have $\zeta_n \in \mathbf{Q}(\zeta_\varpi)$ and therefore $\mathbf{Q}(\zeta_n, \zeta_m) \subset \mathbf{Q}(\zeta_\varpi)$. Conversely, $\varpi/n, \varpi/m$ are coprime so that according to Bézout's identity, there exist integers u, v with

$$u\varpi/n + v\varpi/m = 1.$$

By multiplying by $2\mathrm{I}\pi/\varpi$ and taking the exponential, we find $\zeta_\varpi = \zeta_n^u \zeta_m^v$, proving the reverse inclusion.   □

In general, a "randomly chosen" element of a finite extension is primitive (the reader may try to give a precise meaning to this assertion).

Exercise 5.6.4 If k is not assumed perfect, the primitive element theorem can fail. For example, let $L = \mathbf{F}_p(X, Y)$ be the field of fractions of the polynomial ring $\mathbf{F}_p[X, Y]$. Show that the extension $L(X^{1/p}, Y^{1/p})$ is finite, but is not simple.

We obtain the following fundamental generalization of Proposition 4.6.6.

Corollary 5.6.5 *Let K be a finite extension of k. Then, we have* card $(\mathrm{Hom}_k(K, \Omega)) = [K : k]$.

Proof We write $K = k[x]$ for x primitive and invoke Proposition 4.6.6.   □

Remark 5.6.6 If k is not perfect, the equality is generally false: it is only true for the so-called "separable" extensions of general Galois theory. When k is perfect, all extensions are separable so we will not be bothered by this complication.

Chapter 6
The Galois Correspondence

In this chapter, we prove the main theorem of Galois theory: the Galois correspondence.

We fix a *perfect* field k (Definition 5.4.1) and an algebraically closed field Ω containing it. Recall that any algebraic extension of k is perfect (Proposition 5.4.4).

We have in mind the example $\mathbf{Q} \subset \mathbf{C}$, but also $\mathbf{F}_q \subset \overline{\mathbf{F}}_p$. We will use the fact that every irreducible polynomial P of $k[X]$ is separable (and therefore has deg(P) distinct roots in Ω) and that every finite extension of k is perfect (Proposition 5.4.4).

If $x \in \Omega$ is algebraic over k, we will simply say "conjugates of x" instead of "k-conjugates of x in Ω", that is, the conjugates of x are by definition the roots in Ω of the minimal polynomial P of x over k (Proposition 4.6.6). Since P is irreducible, its roots are simple. But we also know that the map $\sigma \mapsto \sigma(x)$ identifies $\mathrm{Hom}_k(k[x], \Omega)$ and the set of conjugates of x. We therefore have the key formula

Lemma 6.0.1 *The minimal polynomial* P *of an element* $x \in \Omega$ *algebraic over k is*

$$P(X) = \prod_{\sigma \in \mathrm{Hom}_k(k[x],\Omega)} (X - \sigma(x)).$$

We are interested in algebraic extensions K/k, and in fact in finite extensions. According to the theorem of morphism extension (4.9.1), we know that K embeds as a k-algebra in Ω, so it is sufficient to consider the algebraic extensions of k contained in Ω.

© The Author(s), under exclusive license to Springer Nature Switzerland AG 2024
D. Hernandez, Y. Laszlo, *Introduction to Galois Theory*, Springer Undergraduate
Mathematics Series, https://doi.org/10.1007/978-3-031-66182-2_6

6.1   Galois Extensions

Definition 6.1.1 The extension K/k is said to be *Galois* if it is algebraic and if the conjugates of an arbitrary element of K are in K.

Lemma 6.1.2 *Let* K/k *be an algebraic extension of the form* $K = k[x_1, \ldots, x_n]$ *with the* $x_i \in K$. *Then* K/k *is Galois if and only if the conjugates of all the* x_i *over* k *are in* K.

In other words, it is sufficient to verify that the conjugates are in K for a family that generates the extension.

Proof The implication "only if" being obvious, we prove the converse. Let $x \in K$. Then the conjugates of x over k are of the form $\sigma(x)$ with $\sigma \in \mathrm{Hom}_k(K, \Omega)$ where Ω is an algebraic closure of K. But x is of the form $x = P(x_1, \ldots, x_n)$ with $P \in k[X_1, \ldots, X_n]$ a polynomial with n variables. We then have $\sigma(x) = P(\sigma(x_1), \ldots, \sigma(x_n))$. But by hypothesis the $\sigma(x_i)$ are all in K, and therefore $\sigma(x) \in K$.   □

For example, let $k = \mathbf{Q}$ and $K = \mathbf{Q}[2^{1/3}, j]$ with $j = \exp(\frac{2i\pi}{2})$. As $X^2 + X + 1$ (resp. $X^3 - 2$) cancels j (resp. $2^{1/3}$), the conjugates of j are in $\{j, j^2\}$ (resp. $\{2^{1/3}, j2^{1/3}, j^2 2^{1/3}\}$). As these are subsets of K, the previous result implies that K/k is Galois.

We have the following easy but important proposition.

Proposition 6.1.3 *Let* L/k *be a sub-extension of a Galois extension* K/k. *Then* L *is perfect and* K/L *is Galois.*

Proof As K/k is algebraic, so is the extension L/k, so that, since k is perfect, so is K (Proposition 5.4.4). Moreover, the minimal polynomial of $x \in K$ over k (*a fortiori*) has coefficients in L so x is also algebraic over L. Its minimal polynomial over k is divisible by the minimal polynomial of x over L. Therefore, all the L-conjugates of x are also k-conjugates, so they are in K by hypothesis.   □

In general, however, with the same assumptions, L/k is not necessarily Galois (we will see below (Theorem 6.5.1) a necessary and sufficient condition ensuring that L/k is Galois).

For example, let us set $k = \mathbf{Q}$, $\mathrm{K} = \mathbf{Q}[2^{1/3}, j]$ and $\mathrm{L} = \mathbf{Q}[2^{1/3}]$. We saw that K/k is Galois and, according to the previous result, K/L is Galois. We verify however that L/k is not Galois. Indeed, the minimal polynomial of $2^{1/3}$ over k is $X^3 - 2$ (this degree 3 polynomial is irreducible over $\mathbf{Q}$, because we can immediately show that it has no root in $\mathbf{Q}$). So the conjugates of $2^{1/3}$ over $\mathbf{Q}$ are $2^{1/3}$, $2^{1/3} j$ and $2^{1/3} j^2$. But $2^{1/3} j \notin \mathrm{L}$ because $\mathrm{L} \subset \mathbf{R}$.

Recall (Corollary 4.9.5) that the conjugates of $x \in \mathrm{K}$ are also the $\sigma(x)$ with $\sigma \in \mathrm{Hom}_k(\mathrm{K}, \Omega)$. Denote by $\iota : \mathrm{K} \hookrightarrow \Omega$ the inclusion of K in Ω. We have a canonical injective map

$$\iota^* : \mathrm{Aut}_k(\mathrm{K}) \hookrightarrow \mathrm{Hom}_k(\mathrm{K}, \Omega)$$

which associates to an automorphism $\bar{\sigma} \in \mathrm{Aut}_k(\mathrm{K})$

$$\sigma = \iota \circ \bar{\sigma} : \mathrm{K} \xrightarrow{\sigma} \mathrm{K} \xrightarrow{j} \Omega$$

which allows us to identify σ and $\bar{\sigma}$ (and therefore $\mathrm{Aut}_k(\mathrm{K})$ to a subset of $\mathrm{Hom}_k(\mathrm{K}, \Omega)$).

Lemma 6.1.4 *Let K/k be an algebraic extension and $\sigma \in \mathrm{Hom}_k(\mathrm{K}, \Omega)$. Then, $\sigma \in \mathrm{Aut}_k(\mathrm{K})$, i.e. $\sigma(\mathrm{K}) = \mathrm{K}$, if and only if σ leaves K globally invariant, i.e. $\sigma(\mathrm{K}) \subset \mathrm{K}$.*

Proof Indeed, suppose $\sigma(\mathrm{K}) \subset \mathrm{K}$. Let $x_1, \ldots, x_n$ be the n conjugates of $x_1 \in \mathrm{K}$, which are in K by hypothesis. Then, σ leaves $X = \{x_1, \ldots, x_n\}$ globally invariant by hypothesis (since $\sigma(x_i)$ is a conjugate of x_i (Corollary 4.9.5), therefore of x_1 because x_i and x_1 have the same minimal polynomial). As it is a restriction of a morphism of fields, it must be injective. Therefore it induces a bijection of X (since X is finite), so there exists an $x_i \in X$ such that $x_1 = \sigma(x_i)$. As $x_i \in \mathrm{K}$ by hypothesis, σ is surjective. □

We then obtain the following important result.

Corollary 6.1.5 *The inclusion $\mathrm{Aut}_k(\mathrm{K}) \hookrightarrow \mathrm{Hom}_k(\mathrm{K}, \Omega)$ is bijective if and only if K/k is Galois.*

Proof Suppose $\mathrm{Aut}_k(\mathrm{K}) = \mathrm{Hom}_k(\mathrm{K}, \Omega)$. According to Corollary 4.9.5, every conjugate of $x \in \mathrm{K}$ can be written as $\sigma(x)$ for some $\sigma \in \mathrm{Hom}_k(\mathrm{K}, \Omega)$. But $\sigma \in \mathrm{Aut}_k(\mathrm{K})$, so $\sigma(x) \in \mathrm{K}$, proving that K/k is Galois. Conversely, suppose K/k is Galois. Let $\sigma \in \mathrm{Hom}_k(\mathrm{K}, \Omega)$ and $x \in \mathrm{K}$. Then, $\sigma(x)$ is a conjugate of x (Corollary 4.9.5), therefore is in K. As x is arbitrary, we have $\sigma(\mathrm{K}) \subset \mathrm{K}$ and therefore $\sigma \in \mathrm{Aut}_k(\mathrm{K})$ (Lemma 6.1.4). □

Remark 6.1.6 This characterization has the advantage that it actually depends only on K/k and not on Ω: indeed, the conjugates of an algebraic element live in the algebraic closure of k in Ω, which is unique up to isomorphism.

Definition 6.1.7 The *Galois group* of a Galois extension K/k is the group
$$\mathrm{Gal}(K/k) = \mathrm{Aut}_k(K) \overset{(6.1.5)}{=} \mathrm{Hom}_k(K, \Omega).$$

Remark 6.1.8 As the conjugates of $x \in K$ are the $\sigma(x)$, $\sigma \in \mathrm{Hom}_k(K, \Omega)$ (Corollary 4.9.5), if K/k is Galois with Galois group G, the conjugates of x are the $g(x)$, $g \in G$.

We have the following simple example.

Lemma 6.1.9 *The extension* $\mathbf{C}/\mathbf{R}$ *is Galois with Galois group* $\mathrm{Gal}(\mathbf{C}/\mathbf{R}) \simeq \mathbf{Z}/2\mathbf{Z}$ *generated by complex the conjugation.*

Proof We have $\mathbf{C} = \mathbf{R}[\mathrm{I}]$ and since the conjugates of I are I and $-\mathrm{I}$, the extension is Galois. An element of $\mathrm{G} = \mathrm{Gal}(\mathbf{C}/\mathbf{R})$ is uniquely determined by its value at I, which can only be $\bar{\mathrm{I}}$ or $-\mathrm{I}$. Therefore its order $|\mathrm{G}|$ is 1 or 2. But complex conjugation is indeed in G and satisfies $\bar{\mathrm{I}} = -\mathrm{I}$, which implies the result.   □

The following point is easy, but important.

Proposition 6.1.10 *Let* L/k *be a sub-extension of the* Galois *extension* K/k *with* L *perfect. Then,*

(i) $\mathrm{Gal}(\mathrm{K}/\mathrm{L})$ *is a subgroup of* $\mathrm{Gal}(\mathrm{K}/k)$*;*
(ii) *If* L/k *is* Galois, *the restriction of morphisms from* K *to* L *induces (Corollary 6.1.5) a morphism*

$$\mathrm{Gal}(\mathrm{K}/k) \to \mathrm{Gal}(\mathrm{L}/k)$$

which is surjective. Its kernel is $\mathrm{Gal}(\mathrm{K}/\mathrm{L})$*. In other words, we have the exact sequence*

$$\{1\} \to \mathrm{Gal}(\mathrm{K}/\mathrm{L}) \to \mathrm{Gal}(\mathrm{K}/k) \to \mathrm{Gal}(\mathrm{L}/k) \to \{1\}.$$

Proof We know (Proposition 6.1.3) that K/L is Galois. The elements of $\mathrm{Gal}(\mathrm{K}/\mathrm{L})$ are the automorphisms of K that are L-linear while those of $\mathrm{Gal}(\mathrm{K}/k)$ are the automorphisms of K that are k-linear. As L contains k, we deduce an obvious inclusion $\mathrm{Gal}(K/L) \rightarrow \mathrm{Gal}(K/k)$ respecting the composition (and the identity), hence the first point.

According to Corollary 6.1.5, the restriction mapping

$$\mathrm{Hom}_k(K, \Omega) \rightarrow \mathrm{Hom}_k(L, \Omega)$$

is identified with a map

$$\mathrm{Gal}(K/k) \rightarrow \mathrm{Gal}(L/k)$$

which we verify is a morphism. The surjectivity immediately follows from the homomorphism extension theorem (4.9.1). The elements of the kernel are by definition the automorphisms of K fixing L, therefore the elements of $\mathrm{Gal}(K/L)$. The exactness of the sequence in the proposition is only a reformulation of the above, according to Proposition 2.4.3.   □

This will be specified in the Galois correspondence (Theorem 6.5.1) for the case of finite Galois extensions and in Chap. 7 in the general case.

6.2   Characterizations of Galois Extensions

Theorem 6.2.1 *Let* K/k *be a finite extension. Then* K/k *is Galois if and only if the action of* $\mathrm{Aut}_k(K)$ *on the conjugates of any element of* K *is transitive.*

Proof Suppose K/k is Galois and let $x \in K$. According to Corollary 4.9.5, a conjugate y of x is of the form $\sigma(x)$ for $\sigma \in \mathrm{Hom}_k(K, \Omega)$; as $\mathrm{Aut}_k(K) = \mathrm{Hom}_k(K, \Omega)$ (Corollary 6.1.5), the action of $\mathrm{Aut}_k(K)$ on the conjugates of x is indeed transitive. Conversely, let x be a primitive element of K/k (Theorem 5.6.2), thus of degree $[K : k]$. It therefore has $\deg_k(x) = [K : k]$ conjugates and therefore (transitivity), card $\mathrm{Aut}_k(K) \geq [K : k]$. We then invoke again Corollary 6.1.5.   □

The definition of the splitting field of a polynomial was given in Sect. 4.10.

Theorem 6.2.2 *The finite Galois extensions of k are exactly the splitting fields of polynomials.*

Proof Suppose K/k is Galois. According to the primitive element theorem (5.6.2), there exists an x generating K. Let x_i be its conjugates, namely the roots of its minimal polynomial P, which, by hypothesis are in K. We therefore have

$$K = k[x] \subset k[x_i] \subset K$$

and therefore, $K = k[x_i]$ is the splitting field of P, which is what we wanted.

Conversely, suppose $K = k[x_i]$ where the x_i are the roots of a polynomial P. A homomorphism $\sigma \in \mathrm{Hom}_k(K, \Omega)$ permutes the x_i since $P = P^\sigma$. We deduce that it sends $K = k[x_i]$ onto itself so that $\sigma \in \mathrm{Aut}_k(K)$. We then invoke Lemma 6.1.4. □

6.3 The Galois Group of Finite Fields

Let q be the power of a prime number. Recall the results of Sect. 5.2, translated into this new vocabulary:

Proposition 6.3.1 *The extension* $\mathbf{F}_{q^n}/\mathbf{F}_q$ *is Galois, with a cyclic Galois group of order n generated by*

$$F_q : x \mapsto x^q.$$

The subfields of $\mathbf{F}_{q^n}$ *containing* $\mathbf{F}_q$ *are the* $\mathbf{F}_{q^m}$ *with* $m|n$.

In particular, we note that the set of sub-extensions of $\mathbf{F}_{q^n}/\mathbf{F}_q$, that is

$$\{\mathbf{F}_{q^m}/\mathbf{F}_q| \text{ with } m|n\},$$

is in bijection with the set of sub-groups of $\mathrm{Gal}(\mathbf{F}_{q^n}/\mathbf{F}_q) \xrightarrow{\sim} \mathbf{Z}/n\mathbf{Z}$, that is

$$\{((n/m)\mathbf{Z}/n\mathbf{Z}) \simeq \mathbf{Z}/m\mathbf{Z} \text{ with } m|n\}.$$

More precisely, $\mathbf{F}_{q^m}$ is the field of elements of $\mathbf{F}_{q^n}$ fixed by

$$H =< F_q^{m/n} > \xrightarrow{\sim} \mathbf{Z}/m\mathbf{Z}.$$

This is a general phenomenon, as we will now explain.

6.4   Fixed Points

For the rest of this chapter, K/k denotes a finite extension (with as always $K \subset \Omega$ and Ω algebraically closed).

Note that $\mathrm{Aut}_k(K)$ is a group that acts on K.

> **Proposition 6.4.1** *Let K/k be Galois with Galois group G. Then, G has cardinality $[K : k]$ and the space of fixed points K^G of K under the action of G is reduced to k.*

Proof The first point is proven in Corollary 6.1.5. For the second, let x be fixed under the action of G. Its conjugates are of the form $\sigma(x)$ with $\sigma \in$ G according to Theorem 6.2.1, and therefore are equal to x. As the minimal polynomial $P \in k[X]$ of x is separable because it is irreducible (Lemma 5.5.2), it is therefore equal to $X - x$. Thus $x = -P(0) \in k$. $\qquad\square$

Conversely, we prove the following fundamental theorem due to Artin (Fig. 6.1)

> **Theorem 6.4.2 (Artin's Lemma)**   *Let $\mathbf{K}$ be a perfect field and G a finite subgroup of the group of field automorphisms of $\mathbf{K}$. Then, $\mathbf{K}^G$ is perfect and the extension $\mathbf{K}/\mathbf{K}^G$ is finite with Galois group G.*

Proof Let us verify that $\mathbf{K}^G$ is perfect. This only poses a problem in characteristic $p > 0$. Let $x \in \mathbf{K}^G$. Since $\mathbf{K}$ is perfect, it has a p-th root $\xi \in \mathbf{K}$. Since $x = \xi^p$ is invariant under the action of G, we have $\xi^p = g(\xi^p) = g(\xi)^p$ for all $g \in$ G. Since

Fig. 6.1 Emil Artin (1898–1962). Author: Konrad Jacobs and the Mathematisches Forschungsinstitut Oberwolfach. © Konrad Jacobs

the Frobenius morphism is injective, we deduce that ξ is fixed by G and therefore $\xi \in \mathbf{K}^G$, so that $\mathbf{k} = \mathbf{K}^G$ is perfect.

We of course have $G \subset \mathrm{Aut}_{\mathbf{k}}(\mathbf{K})$. Observe that every element x of $\mathbf{K}$ is algebraic of degree $\deg_{\mathbf{k}}(x) \leq \mathrm{card}(Gx)$ and therefore of degree $\leq \mathrm{card}\,G$. Indeed, the polynomial

$$P_x = \prod_{\xi \in Gx} (X - \xi)$$

is invariant under the action of G (that is, all its coefficients are). So P_x is in $\mathbf{k}[X]$ by definition of $\mathbf{k} = \mathbf{K}^G$. Let then $x \in \mathbf{K}$ be an element whose degree over $\mathbf{k}$ is maximal. We then have $\mathbf{K} = \mathbf{k}[x]$. Indeed, otherwise, let $y \in \mathbf{K} - \mathbf{k}[x]$. The extension $\mathbf{k}[x, y]$ is simple (primitive element Theorem 5.6.2), generated by an element z of degree $> \deg_{\mathbf{k}}(x)$, a contradiction. We deduce that

$$[\mathbf{K} : \mathbf{k}] = \deg_{\mathbf{k}}(x) \leq \mathrm{card}\,G.$$

Let Ω be algebraically closed containing $\mathbf{K}$. We then have by Corollary 6.1.5

$$\mathrm{card}\,G \leq \mathrm{card}\,(\mathrm{Aut}_{\mathbf{k}}(\mathbf{K})) \leq \mathrm{card}\,(\mathrm{Hom}_{\mathbf{k}}(\mathbf{K}, \Omega)) = [\mathbf{K} : \mathbf{k}] \leq \mathrm{card}\,G.$$

The result follows thanks to Corollary 6.1.5.   □

6.5   Statement and Proof of the Galois Correspondence

We are now able to prove the main theorem of Galois theory.

Let K/k be a **finite and Galois extension** (contained in Ω) with Galois group G. We recall (Corollary 6.1.5) that we then have

$$G = \mathrm{Hom}_k(K, \Omega).$$

Let $\mathcal{F}$ be the family of subfields L of K containing k, ordered by inclusion. Let $\mathcal{G}$ be the family of subgroups of G, ordered by inclusion. Of course (Proposition 6.1.3), the extension K/L is Galois. We can state the main theorem.

Theorem 6.5.1 (Galois Correspondence) *With the previous notations, we have*

(i) *The map*

$$f : \begin{cases} \mathcal{F} \to & \mathcal{G} \\ L \mapsto \mathrm{Gal}(K/L) \end{cases}$$

is bijective and strictly decreasing, with inverse

$$g : \begin{cases} \mathcal{G} \to \mathcal{F} \\ H \mapsto K^H \end{cases}$$

Now let $H \in \mathcal{G}$.
(ii) *The extension* K/K^H *is Galois with Galois group* H.
(iii) *The restriction mapping*

$$r_H : G = \mathrm{Hom}_k(K, \Omega) \to \mathrm{Hom}_k(K^H, \Omega)$$

identifies the quotient set G/H *with* $\mathrm{Hom}_k(K^H, \Omega)$.
(iv) *The extension* K^H/k *is Galois if and only if* H *is a normal subgroup of* K. *In this case, the previous identification induces an isomorphism*

$$G/H \xrightarrow{\sim} \mathrm{Gal}(K^H/k).$$

(v) *In particular, if* L/k *is Galois, we have a canonical exact sequence*

$$\{1\} \to \mathrm{Gal}(K/L) \to \mathrm{Gal}(K/k) \to \mathrm{Gal}(L/k) \to \{1\}. \tag{6.5.1}$$

Proof We must first verify that we indeed have

$$g(f(L)) = g(\mathrm{Gal}(K/L)) = K^{\mathrm{Gal}(K/L)} = L.$$

As K is Galois over L with Galois group $H = \mathrm{Gal}(L/K)$, we have $K^H = L$ according to Proposition 6.4.1.

Next, we have

$$f(g(H)) = \mathrm{Gal}(K/K^H) \overset{(6.4.2)}{=} H.$$

The two maps f and g are indeed inverses of each other, and in particular are bijective.

It is clear that f is decreasing, its strict character resulting from the bijectivity: we have proved i).

Item ii) is Artin's Lemma (6.4.2).

We now prove item iii). Let H be a subgroup of G. We prove the surjectivity of r_H. Any k-morphism $\sigma_H \in \mathrm{Hom}_k(K^H, \Omega)$ extends to $\sigma \in \mathrm{Hom}_k(K, \Omega)$ according to the theorem of extension of homomorphisms (4.9.1). As K/k is Galois, we have $\sigma(K) = K$ (Lemma 6.1.4), i.e. $\sigma \in G$, so that $r_H(\sigma) = \sigma_H$: the restriction mapping r_H is surjective. Of course, g and gh have the same image if $h \in H$, so we have a surjection

$$\rho_H : G/H \twoheadrightarrow \mathrm{Hom}_k(K^H, \Omega).$$

We then have

$$\mathrm{card}\,\mathrm{Hom}_k(K^H, \Omega) \overset{(5.6.5)}{=} [K^H : k] = [K : k]/[K : K^H]$$

$$= \mathrm{card}\,G/\mathrm{card}\,H = \mathrm{card}(G/H)$$

so that ρ_H is bijective.

Let us prove item iv). Let H be a subgroup of G. Of course, $g \in G$ sends K^H to $K^{gHg^{-1}}$ and therefore g^{-1} sends $K^{gHg^{-1}}$ to K^H, proving

$$g(K^H) = K^{gHg^{-1}}.$$

Suppose K^H/k is Galois. We then have

$$K^H = g(K^H) = K^{gHg^{-1}}$$

and therefore $H = gHg^{-1}$ by the injectivity of the Galois correspondence and therefore $H \lhd G$.

Conversely, if $H \lhd G$, we have

$$g(K^H) = K^{gHg^{-1}} = K^H$$

and K/K^H is Galois.

The group isomorphism is that of Proposition 6.1.10. Now v) clearly follows from the other items. $\square$

Exercise 6.5.2 Let K/k be Galois with group G and K_i/k, $i = 1, 2$, two sub-extensions defined by subgroups G_1, G_2 of subgroups of G. Show that $K_1 K_2$ corresponds to $G_1 \cap G_2$ while $K_1 \cap K_2$ corresponds to the subgroup of G generated by G_1 and G_2 [write these extensions as $\min, \max$ and use that the Galois correspondence is strictly decreasing bijective]. Show that the stabilizer G_x of $x \in K$ in G corresponds to the field $k[x]$ generated by x.

Chapter 7
Addendum: Infinite Galois Correspondence

The consideration of algebraic extensions is fundamental in arithmetic. We will see that the Galois correspondence (Theorem 6.5.1) generalizes as long as we equip the Galois group with a suitable topology, called the profinite topology. In the following, K/k denotes a Galois extension of perfect fields, $\mathcal{F}$ the family of subfields L of K containing k and $\mathcal{F}_{in}$ the subfamily of $\mathcal{F}$ such that L is Galois of finite degree over k. They are ordered by inclusion. We assume the reader is familiar with the basic definitions of general topology.[1]

7.1 Topology of the Galois Group

We start with some reminders from general topology,[2] the details of which the reader will be able to verify without difficulty.

We say that G is a topological group if G is a topological space equipped with a group structure such that the multiplication[3] $\mu : G \times G \to G$ and inverse $G \to G$ operations are continuous. This implies that the right and left multiplications by $g_0 \in G$ are homeomorphisms of G. Therefore, using translations, defining such a topology is equivalent to giving a *basis* $\mathcal{B}$ of neighborhoods of the neutral element $1 \in G$, that is to say a family of subsets of G such that if $H_1, H_2 \in \mathcal{B}$ then there exists an $H \in \mathcal{B}$ such that $H \subset H_1 \cap H_2$. The topology is then separated if, and only if, $\bigcap_{\mathcal{B}} H = \{1\}$. For example, the real or complex linear groups equipped with their topology induced by a norm on the associated matrix space are topological groups.

---

[1] The literature, especially on the Net, is often incomplete or even incorrect on the subject, although going from the finite case to the infinite case is essentially formal. The reader may consult [Bou23] for a slightly different approach.

[2] Cf. [Mun00] for example.

[3] $G \times G$ is equipped with the product topology.

© The Author(s), under exclusive license to Springer Nature Switzerland AG 2024
D. Hernandez, Y. Laszlo, *Introduction to Galois Theory*, Springer Undergraduate
Mathematics Series, https://doi.org/10.1007/978-3-031-66182-2_7

The only topology making finite groups separated is the discrete topology: we will equip them with this topology from now on.

It is not difficult to verify that the topology of the following definition exists:

Definition 7.1.1 The Galois group $G = \text{Gal}(K/k)$ is equipped with the finest topology making the restrictions $G \to \text{Gal}(L/k)$ continuous for $L \in \mathcal{F}_{in}$.

By definition, a basis of open neighborhoods of 1 in G is made up of the kernels $\text{Gal}(K/L)$ (of the restrictions), which are therefore open sets of G since $\text{Gal}(L/k)$ is discrete (and closed[4]).

Since every finite extension is contained in a finite Galois extension, a neighborhood base of 1 in G is $\{G_L = \text{Gal}(K/L), L \in \mathcal{F}_{in}\}$. If $g \in G$, a base of open neighborhoods of g in G is therefore $\{gG_L, L \in \mathcal{F}_{in}\}$. We usefully note

$$gG_L = \{\gamma \in G | \gamma \text{ and } g \text{ coincide on L}\}.$$

Proposition 7.1.2 *The topology of the Galois group satisfies the following properties.*

(1) *The group G is a separated topological group.*
(2) *The restriction homomorphism*

$$\rho = (\rho_L) : G \to \Pi = \prod_{L \in \mathcal{F}_{in}} \text{Gal}(L/k)$$

is an injective homomorphism of topological groups.
(3) *ρ is a homeomorphism onto its image, which is closed in Π. In particular, G is compact.*

Proof To show that the product μ is continuous at (g_1, g_2) is equivalent to showing that the pre-image of a neighborhood $g_1 g_2 G_L$ is a neighborhood of (g_1, g_2), that is, contains an open neighborhood of (g_1, g_2). But

$$g_1 G_L \times g_2 G_L = (G \times g_2 G_L) \cap (g_1 G_L \times G)$$

is such an open set, and for μ the product

$$\mu(g_1 G_L \times g_2 G_L) = g_1 G_L g_2 G_L = g_1 g_2 G_L G_L = g_1 g_2 G_L$$

---

[4] Like any open subgroup, by the way.

because G_L is normal in G. Hence the continuity of μ. The formula

$$(gG_L)^{-1} = G_L g^{-1} = g^{-1} G_L$$

proves the continuity of the inverse and that G is a topological group. The intersection of the G_L is reduced to the identity because every element of K is contained in a finite Galois extension L/k, for example the splitting field in K of its minimal polynomial. So G is separated. Hence (1) is proved.

The separation argument also proves that the restriction homomorphism

$$G \rightarrow \prod_{L \in \mathcal{F}_{in}} \mathrm{Gal}(L/k)$$

is injective. Note that as the topology of each factor of Π is discrete, a neighborhood base of 1 in Π is $\Pi_L = \{\rho_L^{-1}(\mathrm{Id}_L), \ L \in \mathcal{F}_{in}\}$, which is a family of normal subgroups of G. As for G, we deduce that Π is a topological group. By definition of the product topology, ρ is continuous. As we have $\rho(G_L) = \rho(G) \cap \Pi_L$, ρ is open onto its image and therefore is a homeomorphism onto $\rho(G)$.

Let us verify that the image is closed in Π. Let $L_1, L_2 \in \mathcal{F}_{in}$. Let $\underline{\sigma} = (\sigma_L)$ be adherent to the image, so that $\rho(G)$ meets every neighborhood of $\underline{\sigma}$ in Π. In particular, $\rho(G)$ meets the neighborhood $\underline{\sigma} \cdot (\Pi_{L_1} \cap \Pi_{L_2} \cap \Pi_{L_1 \cap L_2})$ of $\underline{\sigma}$. Therefore, there exists a $\sigma \in G$ such that the restrictions of σ to $L_1, L_2, L_1 \cap L_2$ are the corresponding components of $\underline{\sigma}$. In other words, these components coincide on the intersection. This allows us to define $\tilde{\sigma}(x)$ unambiguously for all $x \in K$ by the formula $\sigma_L(x)$ for all $L \in \mathcal{F}_{in}$ such that $x \in L$. We immediately verify that $\tilde{\sigma} \in G$ and that its image is $\underline{\sigma}$. Therefore the image of ρ is closed.

Furthermore, as each $\mathrm{Gal}(L/k)$ is finite and discrete, and therefore compact, Π is compact as a product of compact sets (Tikhonov's theorem) so that $\rho(G)$ is compact. Therefore G, isomorphic to its image, is compact as $\rho(G)$ is closed in a compact set.

□

Remark 7.1.3   The image of the previous morphism is (exercise) the set of compatible families $(g_L)_{L \in \mathcal{F}_{in}}$ of elements of $\mathrm{Gal}(L/k)$, namely such that the images of g_{L_3} in $\mathrm{Gal}(L_i/k)$ is g_{L_i} as soon as $L_3 \in \mathcal{F}$ contains $L_1, L_2 \in \mathcal{F}$. This follows from (one of the) definition(s) of what we call the projective limit of the $\varprojlim_{L \in \mathcal{F}_{in}} \mathrm{Gal}(L/k)$. In this case, it is said to be a profinite group. Note that in a manner analogous to the proof, we can understand the topology of G as follows: $g \in G$ is adherent to a part Γ if for every finite extension L, there exists a $\gamma \in \Gamma$ such that g and γ coincide on L. When k is countable, the number of intermediate extensions K/k is countable, so that G is a compact set in a countable product of discrete sets, therefore is metrizable, like each of them (as a countable product of metric spaces). In the general case, it can be shown that there exist Galois groups whose topology is not induced by a distance.

Exercise 7.1.4 Show that a product of discrete topological spaces is totally disconnected, i.e. its connected components are points. Deduce that Galois groups are totally disconnected. Using that a compact set satisfies the Baire property, show that a Galois group is finite or uncountable.

The following exercise shows that to study the group G, complex (continuous) representations are insufficient. We will need ℓ-adic representations, a whole different story!

Exercise 7.1.5 Let H be a topological group. Show that the continuous group morphisms from H to G identify with compatible families of morphisms (H $\rightarrow$ Gal(L/k))$_{L \in \mathcal{F}_{in}}$ in a sense to be specified.

Exercise 7.1.6 Let N be a norm on $M_d(\mathbf{C})$. It makes $GL_d(\mathbf{C})$ a topological group. Let G = Gal(K/k) as above.

(1) Show that $GL_n(\mathbf{C})$ is open in $M_n(\mathbf{C})$.
(2) Show that the matrix exponential defines a homeomorphism from a neighborhood U of 0 in $M_n(\mathbf{C})$ onto a neighborhood V of Id in $GL_n(\mathbf{C})$.
(3) Show that there exists a neighborhood U' of 0 in U such that $U' + U' \subset U$.
(4) Let M be non-zero in U'. Show that there exists an integer d such that $dM \in U - U'$.
(5) Show that the only subgroup of $GL_d(\mathbf{C})$ contained in exp(U') is {Id}.
(6) Show that every continuous morphism G $\rightarrow$ $GL_d(\mathbf{C})$ factors through the projection G = Gal(K/k) $\rightarrow$ Gal(E/k) for some E $\in \mathcal{F}_{in}$. In particular, its image is finite.

Exercise 7.1.7 For all $n, m \in \mathbf{Z}$, we define $d(n, m) = \sup\{N|$ N! divides $(n - m)\}^{-1}$.

(1) Show that d is a distance and that addition and multiplication of integers are uniformly continuous.
(2) Show that $\{N!\mathbf{Z},\ N \in \mathbf{Z}\}$ is a neighborhood basis of 0.

Let $\hat{\mathbf{Z}}$ be the completion of $(\mathbf{Z}, d)$ and p a prime number. Let G = Gal($\overline{\mathbf{F}}_p/\mathbf{F}_p$) and F the Frobenius morphism.

(3) Show that the map $(\mathbf{Z}, +) \rightarrow$ G that sends 1 to F defines a topological group isomorphism $(\hat{\mathbf{Z}}, +) \simeq$ G.
(4) (For the knowledgeable reader) Show that there exists a canonical isomorphism of topological rings $\hat{\mathbf{Z}} \simeq \prod_{\ell \text{ prime}} \mathbf{Z}_\ell$.

7.2 Infinite Galois Correspondence

> **Theorem 7.2.1 (General Galois Correspondence)** *Let* K/k *be a Galois extension with Galois group* G *and let* $\mathcal{G}$ *be the family of* closed *subgroups of* G.
>
> (i) *The map*
>
> $$f : \begin{cases} \mathcal{F} \to & \mathcal{G} \\ L \mapsto \mathrm{Gal}(K/L) \end{cases}$$
>
> *is bijective and strictly decreasing, with inverse*
>
> $$g : \begin{cases} \mathcal{G} \to \mathcal{F} \\ H \mapsto K^H \end{cases}$$
>
> *Now let* $H \in \mathcal{G}$.
> (ii) *The extension* K/K^H *is Galois with Galois group* H *and the topology of* H *induced by that of* G *coincides with the topology of* H *viewed as a Galois group.*
> (iii) *The extension* K^H/k *is Galois if and only if* H *is a normal subgroup of* K.
> (iv) *If* L/k *is Galois, we have a canonical exact sequence*
>
> $$\{1\} \to \mathrm{Gal}(K/L) \to \mathrm{Gal}(K/k) \to \mathrm{Gal}(L/k) \to \{1\}.$$

Proof Apart from the topological part of (ii), the last three items immediately follow from the Galois correspondence in the finite case (Theorem 6.5.1) and item (i). We have already observed that $\mathrm{Gal}(K/L)$ is closed so that f and g are well defined. We must show that they are inverses of each other.

Let $L \in \mathcal{F}_{in}$ and $x \in K$ be invariant under $\mathrm{Gal}(K/L)$. If $x \notin L$, one of its conjugates $\sigma(x)$, $\sigma \in \mathrm{Hom}_L(K, \Omega)$ is not in L. As K is Galois, $\sigma(K) = K$, and $x \notin K^{\mathrm{Gal}(K/L)}$. Hence $g \circ f = \mathrm{Id}_{\mathcal{G}}$.

To show $f \circ g = \mathrm{Id}_{\mathcal{F}}$ is equivalent to proving the generalized Artin Lemma: if H is a closed subgroup of G, then $H \subset \mathrm{Gal}(K/K^H)$ is an equality and the topology of H coincides with the topology induced by that of G. We let $g \in \mathrm{Gal}(K/K^H)$ and show that g is adherent to H. If E/k is a finite Galois extension of k, we must show the existence of $h \in H$ such that h and g coincide on E. Let x be a primitive element of the extension E/k and set $L = K^H(x)$, the splitting field in K of the minimal polynomial of x over K^H (so that L/K^H is finite Galois). The image H_x by restriction to L of H in $\mathrm{Gal}(L/K^H)$ is therefore a finite group. But L^{H_x} is by definition the set of elements of L fixed by H, so it is $L \cap K^H = K^H$. By the usual Artin Lemma, we have $H_x = \text{Gal}(L/K^H)$. However, $g(x) \in L$ is a K^H-conjugate of x. As the Galois group acts transitively on the conjugates, there exists an $h \in H$ such that $h_x = h_{|L} \in H_x$ satisfies $h_x = g(x)$ so that h and g coincide on E. The coincidence of the topologies is obtained in a similar way.   □

Exercise 7.2.2 Let p be prime and for $n \geq 0$ let K_n be the finite field with p^{2^n} elements (in $\overline{\mathbf{F}}_p$).

(1) Show that K_n is a sequence of increasing fields and that $K = \bigcup_{n \geq 0} K_n$ is a Galois extension of $\mathbf{F}_p$.
(2) Show that the formula $g(x) = x^{p^{2^n - 1}}$ for $x \in K_n$ defines an element of $G = \text{Gal}(K/\mathbf{F}_p)$.
(3) Show $g(x^p) = x$ for all $x \in K$.
(4) Show that the Frobenius morphism $F : x \mapsto x^p$ and g have the same fixed field.
(5) Show that the subgroup of $G = \text{Gal}(K/\mathbf{F}_p)$ generated by F is not closed (one could consider the sequence $F_n = F^{1+p+p^2+\cdots+p^{n-1}}$)
(6) Is the subgroup of $G = \text{Gal}(K/\mathbf{F}_p)$ generated by g closed?

Chapter 8
Cyclotomy and Constructibility

In this chapter, we apply Galois theory to the problem of constructibility of regular polygons. To do this, we first study cyclotomic extensions.

8.1 Cyclotomic Extensions

Let k be a perfect field of characteristic $p \geq 0$ and Ω an algebraic closure of k.

Let $n \geq 1$ be an integer. We also assume that n and p are coprime in the case where $p > 0$. This ensures that $X^n - 1$ and its derivative nX^{n-1} do not have a common root in Ω and therefore that $X^n - 1$ is a separable polynomial (Lemma 5.5.2). The set $\mu_n(\Omega)$ of its roots in Ω is thus a subgroup of Ω^* of finite cardinality n. The group $\mu_n(\Omega)$ is therefore according to Proposition 5.2.1 a cyclic subgroup of Ω^*, isomorphic to $\mathbf{Z}/n\mathbf{Z}$:

$$\mu_n(\Omega) \simeq \mathbf{Z}/n\mathbf{Z}.$$

We recall that, by definition, a primitive n-th root of 1 is a generator of the cyclic group $\mu_n(\Omega)$. Choose such a generator ζ_n. The other primitive n-th roots are the ζ_n^m where m is an integer coprime with n, that is to say,

$$\zeta_n^m, \quad m \in (\mathbf{Z}/n\mathbf{Z})^*,$$

where $(\mathbf{Z}/n\mathbf{Z})^*$ is the multiplicative group of invertible elements of the ring $\mathbf{Z}/n\mathbf{Z}$.[1]

---

[1] We recall that the invertibles of the ring $\mathbf{Z}/n\mathbf{Z}$ are the classes of integers coprime to n (use Bézout's identity to see this).

© The Author(s), under exclusive license to Springer Nature Switzerland AG 2024   89
D. Hernandez, Y. Laszlo, *Introduction to Galois Theory*, Springer Undergraduate
Mathematics Series, https://doi.org/10.1007/978-3-031-66182-2_8

As a result, the extension $k[\zeta_n]$ does not depend on the choice of the primitive root ζ_n.

Definition 8.1.1 An extension of the form $k[\zeta_n]/k$ is called a *cyclotomic extension*.

Proposition 8.1.2 *A cyclotomic extension is Galois.*

Proof As the group $\mu_n(\Omega)$ is generated by ζ_n, the splitting field of the polynomial $X^n - 1$ is simply $k[\zeta_n]$. Thus the extension $k[\zeta_n]/k$ is Galois (Theorem 6.2.2). □

Let

$$G_n = \mathrm{Gal}(k[\zeta_n]/k),$$

the Galois group of the extension $k[\zeta_n]/k$. As the polynomial $X^n - 1$ is of degree n, the cardinality of G_n is less than or equal to n. Later, we will prove a more precise result (Theorem 8.3.10).

8.2 On the Galois Group of the General Cyclotomic Extension

Let us first define the cyclotomic character χ.

Proposition 8.2.1 *There exists a unique map*

$$\chi : G_n \rightarrow (\mathbf{Z}/n\mathbf{Z})^*$$

such that for all $g \in G_n$ and $\zeta \in \mu_n(k)$, we have

$$g(\zeta) = \zeta^{\chi(g)}.$$

Proof The image of ζ_n by an element of the Galois group $g \in G_n$ is uniquely written as

$$g(\zeta_n) = \zeta_n^{\chi(g)} \text{ with } \chi(g) \in \mathbf{Z}/n\mathbf{Z}.$$

As $g(\zeta_n)$ is a primitive root, $\chi(g)$ is invertible in $\mathbf{Z}/n\mathbf{Z}$ (i.e. is the class of an integer coprime to n).

Now, for $\zeta \in \mu_n(k)$, there exists an $m \in \mathbf{Z}/n\mathbf{Z}$ such that $\zeta = \zeta_n^m$. We then have

$$g(\zeta) = g(\zeta_n^m) = (g(\zeta_n))^m = (\zeta_n^{\chi(g)})^m = (\zeta_n^m)^{\chi(g)} = \zeta^{\chi(g)},$$

which proves the existence.

Now suppose there exists a second morphism

$$\chi' : G_n \to (\mathbf{Z}/n\mathbf{Z})^*$$

such that for all $g \in G_n$ and $\zeta \in \mu_n(k)$, we have

$$g(\zeta) = \zeta^{\chi'(g)}.$$

We then have the formula

$$1 = \zeta^{\chi(g)-\chi'(g)}$$

for all $g \in G_n$ and $\zeta \in \mu_n(k)$. By applying it to $\zeta = \zeta_n$, we deduce that $\chi(g)-\chi'(g)$ is divisible by the order of ζ_n in μ_n, therefore by n, so that $\chi(g) - \chi'(g) = 0$.   □

We then obtain:

Proposition 8.2.2 *The cyclotomic character χ defines an isomorphism between the group G_n and the subgroup $\chi(G_n)$ of $(\mathbf{Z}/n\mathbf{Z})^*$. In particular, G_n is commutative.*

Proof Let us first show that the cyclotomic character χ is a group morphism. For $g, g' \in G_n$ and $\zeta \in G_n$, we have

$$(gg')(\zeta) = g(\zeta^{\chi(g')}) = \zeta^{\chi(g)\chi(g')}$$

and therefore $\chi(gg') = \chi(g)\chi(g')$. Now we also have $(gg^{-1})(\zeta) = \zeta = \zeta^{\chi(g)\chi(g^{-1})}$ and therefore $\chi(g^{-1}) = (\chi(g))^{-1}$. As ζ_n generates the extension $k[\zeta_n]$, χ is injective so that χ is an isomorphism onto its image.   □

In the following, we will identify the group G_n and its image $\chi(G_n)$.

8.3   Irreducibility of the Cyclotomic Polynomial Over **Q**

From now on, in the rest of this chapter, $k = \mathbf{Q}$ and $\Omega = \mathbf{C}$.

We can take here $\zeta_n = \exp\left(\frac{2i\pi}{n}\right)$ so that the primitive n-th roots of unity (in **C**) are the complex numbers of the form $\zeta_n^m = \exp\left(\frac{2i\pi m}{n}\right)$, where $m \in (\mathbf{Z}/n\mathbf{Z})^*$.

Definition 8.3.1 We define the *n-th cyclotomic polynomial*

$$\Phi_n(X) = \prod_{m \in (\mathbf{Z}/n\mathbf{Z})^*} \left(X - \exp\left(\frac{2 \mathrm{I} \pi m}{n}\right) \right).$$

An element of G_n, being injective, sends a generator of $\mu_n(\mathbf{C})$ to another generator, and therefore permutes the primitive roots of unity. The coefficients of the cyclotomic polynomial are thus in $\mathbf{Q}[\zeta_n]^{G_n}$ and therefore in $\mathbf{Q}$ according to Lemma 6.4.1. We therefore deduce that $\Phi_n(X)$ is a monic annihilating polynomial of degree

$$\varphi(n) = \mathrm{card}(\mathbf{Z}/n\mathbf{Z})^*$$

of ζ_n in $\mathbf{Q}[X]$.

In fact, we will show that Φ_n is irreducible and has integer coefficients. We start by stating an elementary lemma.

Lemma 8.3.2 (Gauss) *Let* P, Q *be two polynomials with integer coefficients and* Q *monic. Then, the quotient and the remainder of the Euclidean division of* P *by* Q *have integer coefficients.*

Proof It suffices to set up the Euclidean division.   □

We deduce:

Corollary 8.3.3 *We have* $\Phi_n(X) \in \mathbf{Z}[X]$.

Proof Every n-th root of unity has an order d that divides n: it is a primitive d-th root of 1. Conversely, if ζ is a primitive d-th root of 1 with $d \mid n$, it is an n-th root of 1. We deduce that the set of n-th roots of 1 is the disjoint union parameterized by the divisors d of n of the primitive d-th roots. As

$$X^n - 1 = \prod_{\zeta \in \mu_n} (X - \zeta),$$

we deduce the formula

$$X^n - 1 = \prod_{d \mid n} \Phi_d(X). \tag{8.3.1}$$

Starting from $\Phi_1(X) = X - 1 \in \mathbf{Z}[X]$, we obtain by induction on d that Φ_d has integer coefficients according to Lemma 8.3.2, whatever $d \mid n$. This is also true for $d = n$. $\qquad\qquad\qquad\qquad\qquad\qquad\qquad\qquad\qquad\qquad\qquad\qquad\qquad\qquad\qquad\quad$ □

The following lemma is due to Gauss.

Lemma 8.3.4 (Gauss) *Let* $P \in \mathbf{Z}[X]$ *be a non-constant polynomial.*

  (i) *If P is irreducible in* $\mathbf{Z}[X]$, *it is irreducible in* $\mathbf{Q}[X]$.
  (ii) *If P is monic, then the monic irreducible factors of the factorization of* P *in* $\mathbf{Q}[X]$ *have integer coefficients.*

Proof Suppose P is irreducible over **Z** and suppose $P = P_1 P_2$ with $P_1, P_2 \in \mathbf{Q}[X]$ and $\deg(P_1) > 0$. By eliminating the denominators of P_1, P_2, we obtain an identity

$$nP = \overline{P}_1 \overline{P}_2$$

with $n \in \mathbf{Z}$ and $\overline{P}_1, \overline{P}_2 \in \mathbf{Z}[X]$ equal to P_1, P_2 up to scalar multiplication by an element of $\mathbf{N}^*$.

If $n = 1$, we deduce that $\overline{P}_2$ is constant (irreducibility over **Z** of P) and therefore $\deg(P_2) = 0$.

Otherwise, let p be a prime number dividing n. Reducing the relation modulo p, we obtain the identity in $\mathbf{F}_p[X]$

$$0 = (\overline{P}_1 \bmod p)(\overline{P}_2 \bmod p).$$

Since $\mathbf{F}_p[X]$ is a domain ($\mathbf{F}_p$ being a field), we deduce that one of the two polynomials is zero, so that $\overline{P}_1$ or $\overline{P}_2$ has all its coefficients divisible by p. For example, we have $\overline{P}_1 = p\tilde{P}_1$ with $\tilde{P}_1 \in \mathbf{Z}[X]$. By comparing the leading coefficients, we obtain

$$n'P = \tilde{P}_1 \overline{P}_2 \text{ with } n' = \frac{n}{p} \in \mathbf{Z} \text{ and } 1 \le n' < n.$$

Step by step, we arrive at

$$P = P_1^* P_2^*, \text{ with } P_1^*, P_2^* \in \mathbf{Z}[X]$$

and we have reduced to the case $n = 1$.

The proof of the second item is similar. Write $P = P_1 P_2$ with P_1, P_2 monic with rational coefficients. Let n_1, n_2 be the smallest integers > 0 such that $\overline{P}_i = n_i P_i \in$ $\mathbf{Z}[X], i = 1, 2$. If $n_1, n_2 = 1$, the proof is complete. Otherwise, let p be a prime dividing the product $n_1 n_2$. As in the proof of item i), p divides all the coefficients of one of the two polynomials $\overline{P}_i$, say $\overline{P}_1$, and therefore also its leading coefficient, namely n_1. We deduce that we have

$$\frac{n_1}{p} P_1 \in \mathbf{Z}[X],$$

contradicting the minimality of n_1. This concludes the proof.   □

Definition 8.3.5 A complex number is said to be an *algebraic integer* if it is the root of a monic polynomial with integer coefficients.

When the context is clear, we will simply say "integer" instead of algebraic integer.

For example, ζ_n is an integer, but $1/2$ is not (cf. Exercise 8.3.6). We will return to this important concept in Sect. 10.3.

The consistency of the terminology is ensured by the following result.

Exercise 8.3.6 Show that $x \in \mathbf{Q}$ is an integer over $\mathbf{Z}$ if and only if it is in $\mathbf{Z}$.

Gauss's Lemma 8.3.4 immediately gives the following result.

Corollary 8.3.7 *The minimal polynomial of an integer element has integer coefficients.*

Then:

Theorem 8.3.8 *The cyclotomic polynomial Φ_n is irreducible over $\mathbf{Q}$.*

The proof, due to Gauss, is very clever.

Proof Let P be the minimal polynomial of ζ_n. It suffices to prove $\Phi_n | P$, or that all primitive roots of unity cancel P.

Let p be a prime not dividing n and let ζ be a root of P. Then ζ is necessarily a primitive root because $P | \Phi_n$. The key is the following lemma.   □

Lemma 8.3.9 ζ^p *is a root of* P.

Proof Suppose, by contradiction, the opposite. Write

$$X^n - 1 = P(X)S(X)$$

with $S(X) \in \mathbf{Q}[X]$. Since ζ_n is an integer, we have $P(X) \in \mathbf{Z}[X]$ according to Corollary 8.3.7. P(X) being moreover monic, $S(X) \in \mathbf{Z}[X]$. Since $P(\zeta^p)$ is assumed to be non-zero, we have $S(\zeta^p) = 0$. Thus, the polynomials P(X) and $Q(X) = S(X^p)$ have a common complex root. Their GCD (calculated over **Q**) is therefore non-constant, so that P divides Q in $\mathbf{Q}[X]$ (irreducibility of P) and also in $\mathbf{Z}[X]$ since P is moreover monic. Reduce modulo p. We obtain

$$\overline{Q}(X) = \overline{S}(X^p) = (\overline{S}(X))^p$$

using the Frobenius morphism. Since by hypothesis $n \neq 0$ in $\mathbf{F}_p$, $X^n - 1$ and its derivative nX^{n-1} have no common root in $\overline{\mathbf{F}}_p$, so that $X^n - 1$ and $\overline{P}$ have no common factor in $\mathbf{F}_p[X]$. Let Π be an irreducible factor of $\overline{P}$. As it divides $\overline{S}^p$, it divides $\overline{S}$, so that $\Pi^2 | X^n - 1$ in $\mathbf{F}_p[X]$. We obtain a contradiction since $\overline{P}$ is separable.   □

We can now finish the proof of Theorem 8.3.8.

Let then ζ be a root of P and ζ' be any root of Φ_n. We write $\zeta' = \zeta^m$ with $\mathrm{GCD}(m, n) = 1$ (because ζ' is primitive). By decomposing m into a product of prime factors, a repeated application of the lemma gives that ζ' is a root of P and therefore $\Phi_n | P$.   □

Thus,

$$\mathrm{card}(G_n) = [\mathbf{Q}[\zeta_n] : \mathbf{Q}] = \varphi(n),$$

so that χ is an injective morphism (Proposition 8.2.2) between groups of the same cardinality. We have proved the following result.

Theorem 8.3.10 *The cyclotomic character*

$$\chi : \mathrm{Gal}(\mathbf{Q}[\zeta_n]/\mathbf{Q}) \to (\mathbf{Z}/n\mathbf{Z})^*$$

is a group isomorphism.

Exercise 8.3.11 Let K be the cyclotomic extension of **Q** generated by all the roots of unity. Show that K/**Q** is Galois and that its Galois group is $\hat{\mathbf{Z}}^{\times}$ (cf. Exercise 7.1.7).

Remark 8.3.12 It can be demonstrated, but it is significantly above the level of this course, that every abelian extension (Galois with abelian Galois group) of $\mathbf{Q}$ is contained in K: this is the Kronecker–Weber theorem. The general description of abelian extensions is the content of the famous class field theory.

Exercise 8.3.13 Let $n \geq 1$ and p be a prime not dividing n. Show that if Φ_n mod (p) has a root $x \in \mathbf{F}_p$, then x is of order exactly n in $\mathbf{F}_p^*$. Deduce that we have the congruence $p \equiv 1 \bmod (n)$ and then that there are infinitely many prime numbers congruent to 1 modulo n (a weak form of Dirichlet's arithmetic progression theorem).

8.4   Intersections of Cyclotomic Fields

Let d divide n so that $\mathbf{Q}[\zeta_n]$ contains $\mathbf{Q}[\zeta_d]$. The Galois correspondence predicts that $\mathbf{Q}[\zeta_d]$ corresponds to a subgroup of $\mathrm{Gal}(\mathbf{Q}[\zeta_n]/\mathbf{Q})$ of cardinality $\varphi(n)/\varphi(d)$, which must be the kernel of the surjection

$$\mathrm{Gal}(\mathbf{Q}[\zeta_n]/\mathbf{Q}) \twoheadrightarrow \mathrm{Gal}(\mathbf{Q}[\zeta_d]/\mathbf{Q}).$$

Taking into account Theorem 8.3.10, this surjection is nothing other than the canonical group morphism

$$(\mathbf{Z}/n\mathbf{Z})^* \to (\mathbf{Z}/d\mathbf{Z})^*.$$

We find again the statement of Exercise 3.8.3.
Now, let us prove the following result.

Proposition 8.4.1 *For n, m integers, we have*

$$\mathbf{Q}[\zeta_n, \zeta_m] = \mathbf{Q}[\zeta_{\mathrm{LCM}(n,m)}]$$

and

$$\mathbf{Q}[\zeta_n] \cap \mathbf{Q}[\zeta_m] = \mathbf{Q}[\zeta_{\mathrm{GCD}(n,m)}].$$

Proof We set

$$\mathrm{LCM}(n, m) = \pi, \mathrm{GCD}(n, m) = \delta, \mathrm{K} = \mathbf{Q}[\zeta_\pi]$$

and

$$\Gamma_d = \mathrm{Ker}((\mathbf{Z}/\pi\mathbf{Z})^* \to (\mathbf{Z}/d\mathbf{Z})^*)$$

for every d dividing π. We have two subfields $\mathrm{K}_i = \mathbf{Q}[\zeta_n]$, $i = n, m$, of K, defined (Galois correspondence) according to the above by the subgroups Γ_i, $i = n, m$. According to Exercise 6.5.2, we simply need to show $\Gamma_n \cap \Gamma_m = \{1\}$ (which proves $\mathrm{K}_n \mathrm{K}_m = \mathbf{Q}[\zeta_\pi]$) and that the group generated by Γ_n and Γ_m is Γ_δ (proving $\mathrm{K}_n \cap \mathrm{K}_m = \mathbf{Q}[\zeta_\delta]$).

The first point is clear: to say that $g \bmod \pi \in (\mathbf{Z}/\pi\mathbf{Z})^*$ is in the intersection $\Gamma_n \cap \Gamma_m$ is equivalent to saying that n and m divide $g - 1$, in other words $\pi | (g - 1)$.

For the second statement, thanks to the Chinese Lemma, we can assume n and m are powers of p, of the form p^ν, p^μ with for example $0 \le \nu \le \mu$, so that $\delta = p^\nu$. The case where ν or μ is null is trivial. Let us therefore assume $\nu, \mu > 0$. We then have

$$\Gamma_i = 1 + p^i \mathbf{Z}/p^\mu \mathbf{Z}, \ i = \nu, \mu.$$

So the generated group is $\Gamma_\nu = \Gamma_\delta$. $\qquad\qquad\square$

The various extensions are summarized as follows, with here $d = \mathrm{GCD}(n, m)$.

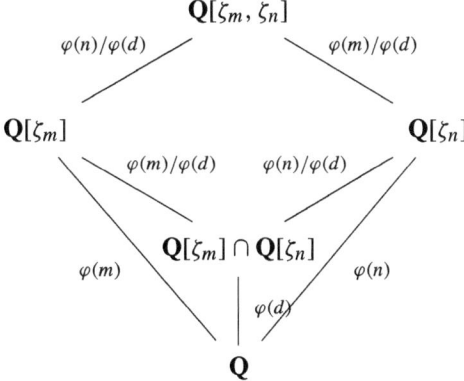

8.5   Constructibility With a Straightedge and Compass

This section describes the application of Galois theory to problems of constructibility with a straightedge and compass. We thus prove the results of Sect. 1.1. First:

Theorem 8.5.1 (Wantzel)   *A complex number z is constructible if and only if there exists a finite sequence of fields $L_0 = \mathbf{Q} \subset L_1 \cdots \subset L_n$ and $[L_{i+1} : L_i] = 2$ with $z \in L_n$.*

Proof Let k be a subfield of $\mathbf{R}$.

First, let us note that a line $D \subset \mathbf{R}^2$ containing two points of k^2 admits an equation of the form

$$ax + by + c = 0 \text{ with } a, b, c \text{ in } k.$$

(We then say that the line is defined over k.) Indeed, if D has an equation of the form $y = Ax + B$ with $A, B \in \mathbf{R}$, we have by hypothesis $x_1 \neq x_2 \in k$ such that $Ax_1 + B, Ax_2 + B \in k$. Therefore $A, B \in k$. Otherwise D has an equation of the form $x = A$. As the line contains a point of k^2, we obtain $A \in k$.

Next, we show that a circle C in $\mathbf{R}^2$ whose center is in k^2 and which contains a point of k^2 admits an equation of the form

$$x^2 + y^2 + ax + by + c = 0 \text{ with } a, b, c \text{ in } k.$$

(We then say that the circle is defined over k.) Indeed, C has an equation

$$(x - x_0)^2 + (y - y_0)^2 = R^2$$

with $(x_0, y_0) \in k^2$ the coordinates of the center. As C contains a point of k^2, we obtain $R^2 \in k$, hence the result.

We now deduce that if a point $(x, y) \in \mathbf{R}^2$ is in the intersection of two lines, of a circle and a line, or of two distinct circles, defined over k, then

$$[k[x] : k] \leq 2 \text{ and } [k[y] : k] \leq 2.$$

For example, in the case of a circle and a line, we write an equation of the line

$$aX + bY + c = 0$$

and of the circle

$$X^2 + Y^2 + dX + eY + f = 0$$

with $a, b, c, d, e, f \in k$. If $b \neq 0$, we write $y = -c/b - ax/b$ and we substitute y into the circle's equation. We obtain an equation of degree at most 2 and $[k[x] : k] \leq 2$. As $y \in k[x]$, we also have $[k[y] : k] \leq 2$. The reasoning is similar if $a \neq 0$. In the case of two circles, we start by subtracting the two circle equations to obtain a relation of the type

$$(d - d')x + (e - e')y + (f - f') = 0.$$

We now prove the theorem. If a point is constructible with a ruler and compass, we obtain the sequence of fields $L_0, \ldots, L_n$ by successively applying the previous results to the points $P_1, P_2, \ldots$.

Conversely, if $[L_i : L_{i-1}] = 2$, we can complete $\{1\}$ to a base $\{1, z\}$ of L_i over L_{i-1}. Then there exist $\lambda, \mu \in L_{i-1}$ such that

$$z^2 = \lambda z + \mu.$$

This can be written as

$$\left(z - \frac{\lambda}{2} \right)^2 = \mu + \frac{\lambda^2}{4}.$$

We set $x = z - \frac{\lambda}{2}$. Then by construction $x^2 \in L_{i-1}^*$. Moreover, as $z \notin L_{i-1}$, we have $x \notin L_{i-1}$ and $L_i = L_{i-1}[x]$. According to Sect. 1.1, x is constructible if the elements of L_{i-1} are. The result therefore follows by induction on n.   □

We then have, according to Theorem 4.2.4

$$[L : Q] = \prod_i [L_{i+1} : L_i] = 2^m.$$

As $Q[z] \subset L_n$, $[Q[z] : Q]$ is a power of 2.

Proposition 8.5.2 *Let $z \in C$ and K be the splitting field of the minimal polynomial of z over Q. z is constructible with a straightedge and compass if and only if $[K : Q]$ is a power of 2.*

Proof Suppose that z is constructible and let $L_0 \subset \cdots \subset L_n$ be a sequence of associated fields. Let z' be a conjugate of z and $\sigma \in \mathrm{Hom}_Q(\overline{Q}, \overline{Q})$ such that $\sigma(z) = z'$. Then

$$\sigma(L_0) \subset \sigma(L_1) \subset \cdots \subset \sigma(L_n)$$

is suitable for z'. So z' is constructible. We have shown that all the conjugates $z_1, \ldots, z_N$ of z are constructible, so the $[Q[z_i] : Q]$ are powers of 2. As

$$K = Q[z_1, \ldots, z_N],$$

the degree $[K : Q]$ divides the product of the $[Q[z_i] : Q]$, and thus it is a power of 2.

Conversely, suppose that $[K : Q]$ is a power of 2. We can apply the Galois correspondence with the Galois extension K/Q whose Galois group has a cardinality that is a power of 2. The result then follows from Exercise 2.7.12. $\square$

> **Theorem 8.5.3 (Gauss–Wantzel)** *The regular polygon with n sides is constructible with a straightedge and compass if and only if n is a product of a power of 2 and distinct Fermat primes.*

Proof The regular polygon with n sides is constructible with a straightedge and compass if and only if $\exp\left(\frac{2I\pi}{n}\right)$ is constructible.

However, we have seen in Theorem 8.3.10 that

$$\mathrm{Gal}\left(Q\left[\exp\left(\frac{2I\pi}{n}\right)\right], Q\right) \simeq (Z/nZ)^*.$$

As $Q[\exp\left(\frac{2I\pi}{n}\right)]/Q$ is Galois, we have

$$\left[Q\left[\exp\left(\frac{2I\pi}{n}\right)\right] : Q\right] = \varphi(n)$$

and according to Proposition 8.5.2, $\exp\left(\frac{2I\pi}{n}\right)$ is constructible if and only if $\varphi(n)$ is a power of 2.

If $\varphi(n)$ is a power of 2, write the factorization of n as a product of prime numbers

$$n = 2^N \prod_{1 \le i \le N} p_i^{N_i}$$

with the p_i odd prime numbers. Then

$$\varphi(n) = 2^{N-1} \prod_{1 \le i \le N} (p_i - 1)p_i^{N_i - 1}.$$

Consequently, all the $N_i = 1$ and the p_i are of the form $p_i = 1 + 2^{M_i}$. We have $M_i = \alpha\beta$ with α odd and β power of 2. We then obtain that $1 + 2^\beta$ divides p_i. The number p_i being prime, we obtain $1 + 2^\beta = p_i$ and thus $M_i = \beta$.

For the converse, if n is of the desired form, it is clear that $\varphi(n)$ is a power of 2. □

Chapter 9
Solvability by Radicals

In this chapter, we study the second historical application of Galois theory: the criterion for solvability by radicals of polynomial equations.

Let k be a perfect field and Ω an algebraic closure of k.

9.1 The Galois Group of a Polynomial

Let P be a non-constant polynomial with coefficients in k and (distinct) roots $x_1, \ldots, x_n$ in Ω. We can assume that P is monic. The roots of P may possibly have multiplicities.

> **Definition 9.1.1** We call the *Galois group of* P *over* k the Galois group $\mathrm{Gal}(P, k) = \mathrm{Gal}(K/k)$ of its splitting field $K = k[x_1, \ldots, x_n]$ (Theorem 6.2.2).

Lemma 9.1.2 *The polynomial* $Q = \prod_{1 \le i \le n}(X - x_i)$ *has coefficients in* k.

This statement is false if we do not assume that k is perfect.

Proof Write $P = \prod_j P_j^{n_j}$ where the P_j are irreducible monic polynomials of $k[X]$ distinct from each other. The field k being perfect, each P_j has simple roots (Theorem 5.5.3). As Q and $\prod_j P_j$ have the same roots and are monic with simple roots, we have $Q \in k[X]$. $\qquad\square$

Remark 9.1.3 By considering $\mathrm{GCD}(P, P')$, the reader will find an efficient algorithm to compute Q without decomposing it into irreducible factors (compare with Sect. 5.3.1).

© The Author(s), under exclusive license to Springer Nature Switzerland AG 2024   103
D. Hernandez, Y. Laszlo, *Introduction to Galois Theory*, Springer Undergraduate
Mathematics Series, https://doi.org/10.1007/978-3-031-66182-2_9

If necessary, by replacing P with Q, we can therefore assume P is **separable**, namely GCD(P, P′) = 1.

The Galois group G of P essentially depends only on P, and not on the choice of the algebraic closure Ω (cf. Theorem 4.9.1, for example).

Recall that, as G leaves k invariant, we have the formula

$$0 = g(P(x_i)) = P(g(x_i)) \text{ for } 1 \le i \le n. \tag{9.1.1}$$

As P has simple roots, there is a unique index $\sigma_g(i)$ such that $x_{\sigma_g(i)} = g(x_i)$. As g is injective, so is σ_g, and therefore $\sigma_g \in S_n$ defines a permutation.

The map $g \mapsto \sigma_g$ defines an action of G on $\{x_1, \ldots, x_n\}$, that is to say a group morphism

$$G \to S_n \text{ where } n = \deg(P).$$

Lemma 9.1.4 *The action of G on the roots of P is faithful.*

Proof It is a matter of showing that the group morphism is *injective*

$$G \hookrightarrow S_n.$$

This is a direct consequence of the fact that the roots generate the splitting field $K = k[x_1, \ldots, x_n]$ as an extension of k.   □

This is the founding viewpoint of Galois and Abel (Fig. 9.1) who saw Galois groups as subgroups of S_n.

Fig. 9.1 Niels Henrik Abel (1802–1829). Author unknown. Source: ETH-Bibliothek Zürich, Bildarchiv, Fel 045631-RE (Source ETH, http://doi.org/10.3932/ethz-a-000161553)

Remark 9.1.5 The reader will easily convince himself that if we change the numbering, we simply conjugate the action by the element of S_n describing the change of numbering. Therefore, more than the permutation, it is its conjugacy class that is well defined, hence its type (Definition 2.6.2).

Proposition 9.1.6 *The polynomial* P *is irreducible* if and only if the action of G *on the roots of* P *is* transitive.

Proof Suppose P is irreducible. The x_i identify with the k-homomorphisms of $k[x_1] = k[X]/(P)$ into Ω (Corollary 4.9.5), which identify with elements of G (as we have seen above). The action is therefore transitive in this case (Theorem 6.2.1). Conversely, suppose the action is transitive and suppose $P = QR$, $Q, R \in k[X]$ with $\deg(Q) > 0$. The formula (9.1.1) applied to Q ensures that the non-empty set of roots of Q is globally invariant under the action of G. As the action of G on the roots of P is transitive, all the roots of P are roots of Q and $Q = P$ (the polynomials P and Q being separable). □

This discussion explains the importance of the symmetric group and its conjugacy classes in the theory. We have the following fundamental example.

Lemma 9.1.7 *Let* $P \in \mathbf{F}_p[X]$ *be an irreducible polynomial of degree* $n > 0$. *Let* $G = \mathrm{Gal}(P, \mathbf{F}_p)$ *and* $F \in G$ *be the Frobenius morphism. Then* G *is a cyclic subgroup of* S_d, F *is a cycle of length* d *and* $\mathrm{card}(G) = d$.

Proof Since P is irreducible and $\mathbf{F}_p$ is perfect, the polynomial P is separable. Let $(z_j)_{1 \le j \le d}$ be the roots of P in $\overline{\mathbf{F}}_p$. Since G acts on $(z_j)_{1 \le j \le d}$, we can embed G in S_d, where we identify S_d with $\mathrm{Bij}(z_1, \ldots, z_d)$. Since P is an irreducible polynomial, P is the minimal polynomial of z_1 over $\mathbf{F}_p$ and its roots are the conjugates of z_1. The field $\mathbf{F}_p$ and the splitting field of P being finite, G is a cyclic group generated by F (Theorem 5.2.3). The conjugates of z_1 are therefore exactly the $F^n(z_1)$, $n = 0, \ldots, d-1$. The image of F in S_d is thus the cycle $(z_1, F(z_1), \ldots, F^{d-1}(z_1))$. It is a cycle of length d, which completes the proof. □

9.2 The Discriminant

Let $G \subset S_n$ be the Galois group of a polynomial $P \in k[X]$. Here we give a simple condition which allows us to decide if we have $G \subset A_n$.

Proposition 9.2.1 *The element*

$$\mathrm{disc}(P) = (-1)^{\frac{n(n-1)}{2}} \prod_{\substack{x \neq y \\ P(x)=P(y)=0}} (x - y)$$

is an element of k^*.

In the case where k is of characteristic different from 2, $\mathrm{disc}(P)$ is a square in k^* if and only if $G \subset A_n$.

We say that $\mathrm{disc}(P)$ is the discriminant of P. This is a particular case of the notion of *resolvent*.

Proof Clearly $\mathrm{disc}(P)$ is non-zero and invariant under G, so (Proposition 6.4.1) is in k^*. Choose an ordering on the roots of P, which we write as $x_1, \ldots, x_n$. We then have

$$\mathrm{disc}(P) = (-1)^{\frac{n(n-1)}{2}} \prod_{i \neq j} (x_i - x_j).$$

Let

$$\sqrt{d} = \prod_{i < j} (x_i - x_j) \in \Omega.$$

We have

$$\sqrt{d}^2 = \mathrm{disc}(P).$$

Let S_n operate on the indices $\{1, \ldots, n\}$. We then have for $\sigma \in S_n$

$$\sigma(\sqrt{d}) = \prod_{i < j} (x_{\sigma(i)} - x_{\sigma(j)}).$$

The pairs $(\sigma(i), \sigma(j))$ are of two types: either $\sigma(i) < \sigma(j)$, and we find a factor of the initial product, or $\sigma(i) > \sigma(j)$ and we find the negative of a factor of the initial product. We therefore have

$$\sigma(\sqrt{d}) = (-1)^{|\sigma|} \prod_{i < j} (x_i - x_j)$$

where

$$|\sigma| = \text{card}\{(i, j) \text{ such that } i < j \text{ and } \sigma(i) > \sigma(j)\}.$$

We then remember the formula (Sect. 2.6)

$$\epsilon(\sigma) = (-1)^{|\sigma|}$$

so that

$$\sigma(\sqrt{d}) = \epsilon(\sigma)\sqrt{d}.$$

Therefore, if $G \subset A_n$, we certainly have $\sqrt{d} \in k$, whatever the characteristic of k.

Conversely, if $\sqrt{d} \in k^*$, we have

$$\sqrt{d} = \epsilon(\sigma_g)\sqrt{d}.$$

So $\epsilon(\sigma_g) = 1$ in k, which means $\epsilon(\sigma_g) = 1$ in $\mathbf{Z}$ if the characteristic of k does not divide $1 + \epsilon(g) = 2$. Hence the lemma.   □

Exercise 9.2.2 Show that the Galois group of a polynomial of degree 2 is trivial or isomorphic to $\mathbf{Z}/2\mathbf{Z}$. Show that in degree 3, characteristic different from 2, the Galois group is isomorphic to $\mathbf{Z}/3\mathbf{Z}$ or S_3. Show that the latter case only occurs if disc(P) is not a square unless P has a root in k. Deduce that the Galois group of $X^3 - 2$ over $\mathbf{Q}$ is S_3 (isomorphic to the dihedral group D_6).

Exercise 9.2.3 Let k be a perfect field of characteristic $p \geq 0$. Calculate the discriminant of $P = X^n - 1$. Show that P is separable if and only if p does not divide n, which we now assume. Let K be the splitting field of P (in an algebraic closure $\bar{k}$ of k). Give a necessary and sufficient condition on n for the action of $\text{Gal}(K/k)$ on $\mu_n(\bar{k})$ to be in A_n, at least if $p \neq 2$.

Exercise 9.2.4 Let P be a separable polynomial with coefficients in k of characteristic 2. We denote by x_i its roots in $\bar{k}$ and G the Galois group of $k[x_i]/k$, which therefore acts on the x_i, and thus embeds in S_n. Show that $x_i + x_j$ and $x_i^2 + x_j^2$ are non-zero if $i < j$. Show that $a = \sum_{i<j} \frac{x_i x_j}{x_i^2 + x_j^2}$ is an element of k. Let $b = \sum_{i<j} \frac{x_i}{x_i + x_j} \in \bar{k}$. Show that on the one hand $b^2 + b = a$ and, on the other hand, $g(b) = b$ or $b + 1$ depending on whether $g \in A_n$ or $g \notin A_n$. Deduce that we have $G \subset A_n$ if and only if a is of the form $x^2 + x$ with $x \in k$.

9.3 Cyclic Extensions

In this section, n denotes an integer greater than or equal to 2 and k is a perfect field such that $\mu_n(k)$ has cardinality n.

We then abusively say that "k contains the n-th roots of unity". In particular, the characteristic of k does not divide n (but this is obviously not sufficient in general).

As any finite subgroup of k^* is cyclic, we then know that the group $\mu_n(k)$ is isomorphic to $\mathbf{Z}/n\mathbf{Z}$ (not canonically in general). Thus, there exist in k the primitive roots (of order n) of 1 (precisely, there are $\varphi(n)$ primitive roots).

Definition 9.3.1 A *cyclic extension* is an extension K/k with K the rupture field of a polynomial of the form $P(X) = X^n - a$ with $a \in k \setminus \{0\}$.

Note that then $K = k[\alpha]$ with $\alpha = a^{1/n}$ a root of P.

Lemma 9.3.2 *A cyclic extension is Galois.*

Proof The roots of $X^n - a$ are the multiples of α by the n-th roots of unity. By hypothesis these latter are in k. Thus, K is the splitting field of $X^n - a$, and therefore is Galois over k.   □

Let $G = \mathrm{Gal}(K/k)$ be the Galois group of the extension K/k.

Lemma 9.3.3 *There exists a unique group morphism*

$$\kappa : G \to \mu_n(k)$$

such that $\kappa(g) = g(\alpha)/\alpha$ for $g \in G$.

Proof If $g \in G$, the element $g(\alpha)$ is a root of P, therefore of the form $\zeta \alpha$ for $\zeta \in \mu_n(k)$. The map κ is therefore well defined. Now, we have $g(\alpha) = \kappa(g)\alpha$. By composing with g^{-1} we get

$$\alpha = g^{-1}(\kappa(g))g^{-1}(\alpha).$$

But $\kappa(g) \in \mu_n(k) \subset k$, so $g^{-1}(\kappa(g)) = \kappa(g)$. Thus

$$\alpha(\kappa(g))^{-1} = g^{-1}(\alpha) \text{ and } \kappa(g^{-1}) = (\kappa(g))^{-1}.$$

Let now $h \in G$. We have similarly $g(\kappa(h)) = \kappa(h)$ and

$$(gh)(\alpha) = g(\kappa(h)\alpha) = g(\kappa(h))g(\alpha) = \kappa(h)\kappa(g)\alpha.$$

We thus obtain $\kappa(gh) = \kappa(g)\kappa(h)$. We have shown that κ is a group morphism.   □

Lemma 9.3.4 κ *is injective and* G *is cyclic of cardinality* d *dividing* n. *Moreover, the following assertions are equivalent:*

(i) P *irreducible,*
(ii) a *is not a* d-*th power in* k *for any divisor* d *of* n *distinct from* 1,
(iii) $G = \mu_n(k)$.

Proof The injectivity of κ is clear. As $\mu_n(k)$ is cyclic of cardinality n, the order of G is a divisor δ of n. To say that P is irreducible is to say that $[K : k] = n$ and therefore that κ is surjective. Suppose that P is irreducible, and therefore that $G = \mu_n(k)$. Let $\zeta = \kappa(g)$ be primitive in $\mu_n(k)$ and $d \,|\, n$ such that $\alpha^d \in k$ (that is, $a \in k^{\frac{n}{d}}$). We have $g(\alpha^d) = \zeta^d \alpha^d$ but also $g(\alpha^d) = \alpha^d$ because $\alpha^d \in k$. We therefore have $\zeta^d = 1$ and therefore $n|d$, so $d = n$. Conversely, if P is not irreducible, the cardinality δ of G strictly divides n. For all $g \in G$, we have $g(\alpha)/\alpha \in \mu_\delta(k)$ and therefore $g(\alpha^\delta) = \alpha^\delta$. Therefore $\alpha^\delta \in k$. □

The following converse is, in a sense, quite surprising.

We recall that the cardinality of $\mu_n(k)$ is n (otherwise the following theorem, due to Kummer (Fig. 9.2), is not valid).

Theorem 9.3.5 (Kummer) *Let* K/k *be a Galois extension with Galois group* $G = \mathrm{Gal}(K/k)$ *cyclic of order* n. *Then, the extension* K/k *is cyclic.*

Proof Let g be a generator of the Galois group: it satisfies $g^n = \mathrm{Id}$, regarded in $\mathrm{End}_k(K)$. We know that K is a vector space of degree n over k (Proposition 6.4.1). By hypothesis, $X^n - 1$ is split over k with simple roots, so g is diagonalizable. The formula

$$g(xy^{-1}) = g(x)g(y)^{-1}$$

Fig. 9.2 Ernst Kummer (1810–1893). Author unknown. Source: Wikimedia Commons

ensures that the set of eigenvalues of G is a subgroup of $\mu_n(k)$ and therefore is a cyclic group of order d. If we had $d < n$, we would have $g^d = \text{Id}$, which is not the case because g is a generator and therefore there is an eigenvalue of g which is a primitive n-th root ζ of 1. Let x be a corresponding (non-zero) eigenvector. By construction, x has at least n conjugates. These are the $(\zeta^i x)_{1 \le i \le n}$ which are distinct and necessarily in K. The extension K/k is Galois so that $K = k[x]$. Thus, the $(\zeta^i x)_{1 \le i \le n}$ are all the conjugates of x, and therefore the minimal polynomial of x is

$$\prod_{1 \le i \le n} (X - \zeta^i x) = X^n - a$$

with $a \in k$.   □

Remark 9.3.6 If we read the previous proof carefully, the underlying phenomenon is the following. Denote by K_a the eigenspace $\text{Ker}(g - a\text{Id})$. We then have

$$K = \oplus_{1 \le i \le n} K_i \text{ with } K_i = kx^i, \tag{9.3.1}$$

where x is non-zero in K_1.

Exercise 9.3.7 Suppose n is coprime to the characteristic of k. Let $a \in k$. Show that $P = X^n - a$ is separable and that its Galois group G is an extension of two abelian groups, in other words that we have an exact sequence $1 \to G_1 \to G \to G_2 \to 1$ with G_1, G_2 abelian (and even G_2 cyclic). Give an example where G is not commutative.

Of course, any sub-extension of a cyclotomic extension $\mathbf{Q}[\zeta_n]$ has an abelian Galois group. A difficult theorem, known as the Kronecker–Weber Theorem (Figs. 9.3 and 9.4), ensures that the converse is true! This is a consequence of the vast class field theory, which naturally leads to the visionary theory of Langlands (Fig. 9.5), a very difficult and very active subject at the moment.

Fig. 9.3 Leopold Kronecker
(1823–1891). Author
unknown. Source: Wikimedia
Commons

Fig. 9.4 Heinrich Martin
Weber (1842–1913). Author
unknown. Source:
ETH-Bibliothek Zürich,
Bildarchiv, Dia 326-741

Fig. 9.5 Robert Langlands
(1936–). © Dan Komoda,
Institute for Advanced Study

9.4 Applications to Polynomial Equations

In this section, we assume that k has characteristic zero.[1]

> **Definition 9.4.1** We say that a field extension K/k is *radical* if there exists a
> sequence of fields K_i, $i = 0, \ldots, n$, such that
>
> $$k = K_0 \subset \cdots \subset K_n = K$$
>
> and $K_{i+1} = K_i[x_i]$ and a certain power of x_i is in K_i.
>   We say that K/k is *solvable* if there exists a finite extension L/K containing
> K such that L/k is radical.

We also say that K/k is "solvable by radicals" when it is solvable.
  Thus, if K/k is solvable, every element $x \in K$ can be expressed using rational
fractions and successive radicals from elements of k.

---

[1] But no assumption about $\mu_n(k)$ is made.

Therefore, to say that the splitting field of $P \in k[X]$ is solvable (over k) is to say that its roots are expressed rationally from successive extractions of elements of k: this is the intuitive notion of solvability by radicals!

Theorem 9.4.2 (Galois) *Let K/k be Galois. If K/k is solvable, then $G = \mathrm{Gal}(K/k)$ is solvable.*

In fact, the converse of the theorem is also true.[2] Let us move on to the proof of the theorem.

Proof By hypothesis, K is contained in L with L/k radical. We therefore have a sequence of fields

$$k = \overline{L}_0 \subset \overline{L}_1 \cdots \subset \overline{L}_n = L$$

such that $\overline{L}_{i+1} = \overline{L}_i[x_i]$ and $x_i^{n_i} \in \overline{L}_i$.

The problem is that the $\overline{L}_i$ have no reason to be Galois over k. Let us remedy this. We want to use Kummer's theorem. Let therefore n be a multiple of all the n_i and X_i the set of conjugates (over k) of the x_j, $0 \le j \le i$.

We then set

$$L_{i+1} = k[\zeta_n, X_i], i = 0, \ldots, n-1,$$

which by construction is Galois over k (as before, ζ_n denotes a primitive nth root of unity in Ω). We set $L_{-1} = k, X_{-1} = \{\zeta_n\}$ so that we have

$$L_{i+1} = L_i[X_i] \text{ for } i \ge -1$$

with L_i Galois over $L_{-1} = k$ for all i and therefore *a fortiori* over L_j, $-1 \le j \le i$.

As $\mathrm{Gal}(K/k)$ is a quotient (Theorem 6.5.1) of $\mathrm{Gal}(L_n/L_{-1})$, it is enough to show that the latter is solvable (Proposition 2.7.6). Let us show by induction on i that $\mathrm{Gal}(L_i/L_{-1})$ is solvable. □

Lemma 9.4.3 *Each group $\mathrm{Gal}(L_{i+1}/L_i)$, $i \ge -1$ is solvable.*

Proof As $\mathrm{Gal}(L_0/L_{-1})$ is commutative (Proposition 8.2.2), we can assume $i \ge 0$.

If $i \ge 0$, L_{i+1} is obtained by "adding" to L_i the conjugates $(y_j)_{1 \le j \le \deg_k(x_i)}$ of x_i over k. We therefore have a tower of extensions

$$M_0 = L_i \subset M_1 = L_i[y_1] \subset \cdots \subset M_d = L_i[y_1, \ldots, y_d] = L_{i+1}.$$

2 This is not difficult to prove if we know Kummer's theory.

Since $y_j^n \in L_i$ and $\zeta_n \subset L_i$, all L_i-conjugates of y_j, $j \leq \delta$ are in M_δ, which is therefore Galois over M_0, and also over the $M_{\delta'}$, $\delta' \leq \delta$. Moreover, $M_{\delta+1} = M_\delta[y_{\delta+1}]$, and each intermediate elementary extension $M_{\delta+1}/M_\delta$ is cyclic, hence Galois with a cyclic Galois group. We have

$$1 = G_0 = \mathrm{Gal}(M_d/M_d) \subset G_2 = \mathrm{Gal}(M_d/M_{d-1}) \subset \cdots \subset G_d = \mathrm{Gal}(M_d/M_0).$$

We also have

$$G_{\delta+1}/G_\delta \xrightarrow{\sim} \mathrm{Gal}(M_{\delta+1}/M_\delta)$$

(thanks to the fundamental exact sequence (6.5.1)) and therefore $G_{\delta+1}/G_\delta$ is abelian. Proposition 2.7.6 ensures that G_d is solvable.   □

We can now complete the proof of the theorem.

The Galois theory (6.5.1) gives us exact sequences

$$1 \to \mathrm{Gal}(L_{i+1}/L_i) \to \mathrm{Gal}(L_{i+1}/L_{-1}) \to \mathrm{Gal}(L_i/L_{-1}) \to 1.$$

The result follows thanks to the lemma and to Proposition 2.7.6.   □

Let us move on to the application to equations: can we solve a polynomial equation by radicals, that is, can we express its roots in terms of its coefficients? This amounts to showing that the extension generated by the roots is solvable over the field generated by the coefficients.

Let $L = \mathrm{Frac}(\mathbf{C}[X_1, \ldots, X_n])$ be the field of fractions of $\mathbf{C}[X_1, \ldots, X_n]$, where the X_i are indeterminates. The symmetric group S_n acts on L by permutation of the indices. Let $K = L^{S_n}$: the extension L/K is Galois with Galois group S_n according to Artin's Lemma. We also know (Sect. 11.5) that we have

$$K = \mathrm{Frac}(\mathbf{C}[\sigma_1, \ldots, \sigma_n]),$$

where the σ_i are the elementary symmetric polynomials in the X_i defined by the identity

$$\prod_{i=1}^{n}(X - X_i) = X^n + \sum_{i=1}^{n}(-1)^i \sigma_i X^{n-i}. \tag{9.4.1}$$

Remark 9.4.4 We can also invoke Artin's Lemma and the trivial upper bound $[L : \mathrm{Frac}(\mathbf{C}[\sigma_1, \ldots, \sigma_n])] \leq n!$ to prove that $K = \mathrm{Frac}(\mathbf{C}[\sigma_1, \ldots, \sigma_n])$.

The formula (9.4.1) also proves that L is the splitting field over K of

$$P(X) = X^n + \sum_{i=1}^{n-1}(-1)^i \sigma_i X^{n-i}.$$

The general polynomial equation with coefficients in K is

$$X^n + \sum_{i=1}^{n} (-1)^i \sigma_i X^{n-i} = 0.$$

To say that this equation is solvable by radicals means that L/K is solvable. This is not the case if $n \geq 5$:

Theorem 9.4.5 (Abel, Galois)   L/K *is not solvable as soon as* $n \geq 5$.

Proof It suffices to use Theorem 9.4.2 and Proposition 2.7.10.   □

Chapter 10
Reduction Modulo p

In this chapter, we will provide methods for studying Galois groups of monic polynomials P with integer coefficients by reduction modulo p (p prime number), that is using the polynomial $\overline{P} \in \mathbf{F}_p[X]$ obtained by reduction modulo p of the coefficients of P. This will allow us to reduce to the situation of the field $\mathbf{F}_p$, which is, at least theoretically, simpler: for example, we know how to factorize polynomials into irreducible polynomials (see Sect. 5.3, Berlekamp's algorithm (Fig. 10.1)), the extensions are always Galois with cyclic Galois groups, and we have the bonus of a canonical generator, the Frobenius morphism.

So we are given a prime number p. In the case where $\overline{P}$ is separable, we will compare the Galois group Gal(P, $\mathbf{Q}$) of P (i.e. that of its splitting field (Sect. 9.1)) and that of $\overline{P}$, namely Gal($\overline{P}$, $\mathbf{F}_p$).

The main result for us is that, **under these conditions, there exists an element of** Gal(P, $\mathbf{Q}$), **unique up to conjugation, whose conjugacy class in the symmetric group** $S_{\deg(P)}$ **is the same as that of the Frobenius morphism in** Gal($\overline{P}$, $\mathbf{F}_p$) (for the canonical embeddings of Galois groups in the symmetric groups).

Let us start by stating the result in a form that we will use most in practice.

10.1 Theorem of Reduction Modulo p

We recall that the Galois group of a polynomial admits a canonical embedding into the group of permutations of the roots of this polynomial. We have seen that if this polynomial is separable of degree n, each numbering of the roots defines an embedding of its Galois group to S_n which is well defined up to conjugation in S_n.

Let P be a *monic* separable polynomial with integer coefficients of degree n. Let p be a prime number and $\overline{P}$ the image of P in $\mathbf{F}_p[X]$. As P is monic, the degree of $\overline{P}$ is n.

© The Author(s), under exclusive license to Springer Nature Switzerland AG 2024
D. Hernandez, Y. Laszlo, *Introduction to Galois Theory*, Springer Undergraduate
Mathematics Series, https://doi.org/10.1007/978-3-031-66182-2_10

Fig. 10.1 Elwyn Berlekamp (1940–). Photographer: George M. Bergman. © George M. Bergman

> **Theorem 10.1.1 (Of Reduction Modulo p)** *Suppose that $\overline{P}$ is separable. Then* $\mathrm{Gal}(P, \mathbf{Q})$ *admits a subgroup isomorphic to* $\mathrm{Gal}(\overline{P}, \mathbf{F}_p)$. *For any permutation in* $\mathrm{Gal}(\overline{P}, \mathbf{F}_p) \subset S_n$, *there exists a permutation in* $\mathrm{Gal}(P, \mathbf{Q}) \subset S_n$ *of the same type. In particular, if $\overline{P}$ is irreducible, there exists a cycle of length n in* $\mathrm{Gal}(P, \mathbf{Q}) \subset S_n$.

The reader may leave the proofs for a second reading (although they are not difficult).

10.2   Specialization of the Galois Group

Let P be a *monic* separable polynomial with integer coefficients.[1] Let

$$A = \mathbf{Z}[z_1, \ldots, z_n]$$

denote the subring of $\mathbf{C}$ generated by the complex roots $z_1, \ldots, z_n$ of P. By construction, all the z_i are integers (Definition 8.3.5).

Lemma 10.2.1 *The field of fractions K of A is the splitting field of P in* $\mathbf{C}$.

---

[1] The reader can adapt the proof to the case where the coefficient ring $\mathbf{Z}$ is replaced by a factorial ring, or even an integrally closed ring.

Proof Indeed, A is contained in the splitting field L of P because $z_i \in$ L for all $i \in \{1, \ldots, n\}$, and therefore K is contained in L. Moreover, since P is split over K, we have K $=$ L by the minimality of L. □

The fundamental observation is that all elements of A are integers. To demonstrate this, we will provide a characterization of integers in every way analogous to that of algebraic numbers (Proposition 4.6.1).

10.3 Sums and Products of Integers

Let C be a ring (in general C is not **C**!).

Definition 10.3.1 Let B be a C-algebra. We say that $b \in$ B is an *integer over* C if b nullifies a *monic* polynomial with coefficients in C.

When B is a subring of **C**, regarded as an algebra over C $=$ **Z**, we find the notion of algebraic integer (Definition 8.3.5). The following proposition generalizes Proposition 4.6.1. One can refer to Sect. 3.7 for the definitions of modules over a ring.

Proposition 10.3.2 *Let* B *be a* C*-algebra and* $b \in$ B. *Then,* b *is an integer over* C *if and only if* b *is contained in a sub-ring* B$'$ *of* B *which is a* C*-module of finite type.*

We recall (Sect. 3.7) that B$'$ is said to be a C-*module of finite type* if there exists a finite family b_i of elements of B$'$ such that every element of B$'$ is a linear combination of the b_i with coefficients in C.

Proof The direct part is clear: if b has a monic annihilating polynomial of degree d, then C $=$ C[b] is generated by $1, \ldots, b^{d-1}$. Conversely, suppose $b \in$ B$'$ of finite type over C, generated by $b_1, \ldots, b_n$. There exist $c_{i,j} \in$ C such that for all j

$$bb_j = \sum_i c_{i,j} b_i$$

($\alpha = (c_{i,j})$ is a matrix of the homothety $h_b \in$ End$_C$(B$'$) of ratio b in C). Let P $=$ det(XId $- \alpha$) be the characteristic polynomial of α: it is a monic polynomial of C[X] that annihilates α (Cayley–Hamilton theorem) and therefore *a fortiori* h_b. But, we have $0 = P(h_b) \cdot 1 = P(b)$, which is what we wanted. □

As in Proposition 4.6.7, we deduce

Corollary 10.3.3 *The set of elements of* B *that are integers over* C *is a subring of* B.

Proof Indeed, if $x, y \in$ B are integers over C, let's say annihilated by monic polynomials with coefficients in C of degree n, m, then both $x - y$ and xy are contained in $C[x, y]$. But $C[x, y]$ is generated by the monomials $x^i y^j, 0 \leq i \leq n, 0 \leq j \leq m - 1$ and therefore is of finite type over C.   □

Corollary 10.3.4 *The set of algebraic integers is a subring of* C.

In particular, all elements of A are integers.

Exercise 10.3.5 (Kronecker's Lemma)   Let z be a complex number which is an integer over **Q** and let $(z_i)_i$ be its conjugates. Show that the z_i are integers. Show that all polynomials $\prod_i (X - z_i^n)$ where $n \in$ **N** have integer coefficients. From this, deduce that if $|z_i| \leq 1$ for all i, then either $z = 0$, or the z_i are roots of unity.

10.4   Norm of the Elements of A

For every algebraic complex number z over **Q**, we define its norm $N(z)$ as the product of its complex conjugates. If P is the minimal polynomial of z, we obviously have the formula

$$N(z) = (-1)^{\deg(P)} P(0)$$

so that

$$N(z) \in \mathbf{Q}.$$

For example, $N(z) = z$ if $z \in$ **Q** while $N(\sqrt{2}) = -2$. If z is an integer, we therefore have $N(z) \in$ **Z** since $P \in$ **Z**[X] (Corollary 8.3.7).

Lemma 10.4.1 *The quotient ring* $\overline{A} = A/pA$ *is non-zero.*

Proof Suppose the contrary. We would then have $1 = pa$ for some $a \in$ A. However, the $d = \text{card}(\text{Hom}_{\mathbf{Q}}(\mathbf{Q}[a], \mathbf{C}))$ distinct conjugates of a are the complex numbers $\sigma(a)$ with $\sigma \in \text{Hom}_{\mathbf{Q}}(\mathbf{Q}[a], \mathbf{C})$. As $\mathbf{Q}[a] = \mathbf{Q}[pa]$, the complex number pa has d

distinct conjugates which are the $p\sigma(a)$ with $\sigma \in \text{Hom}_{\mathbf{Q}}(\mathbf{Q}[a], \mathbf{C})$. We deduce the formula

$$N(pa) = p^{\deg_{\mathbf{Q}}(a)} N(a)$$

on the one hand, and, on the other hand,

$$N(pa) = N(1) = 1.$$

This is absurd because $N(z) \in \mathbf{Z}$, a being an algebraic integer according to Corollary 10.3.4. $\qquad\qquad\square$

10.5 Decomposition Groups

Let $\bar{\mathfrak{p}}$ be a maximal ideal of the (non-zero!) ring $\overline{A}$.[2] Then $k = \overline{A}/\bar{\mathfrak{p}}$ is a field. Let $\mathfrak{p}$ be the pre-image of $\bar{\mathfrak{p}}$ in A, in other words the kernel of the canonical surjection

$$A \twoheadrightarrow \overline{A} \twoheadrightarrow k.$$

As the image of p is zero in $\overline{A} = A/pA$, the field k is of characteristic p.

Remark 10.5.1 It is useful to observe that we have $\mathfrak{p} \cap \mathbf{Z} = p\mathbf{Z}$. Indeed, $\mathfrak{p} \cap \mathbf{Z}/p\mathbf{Z}$ is the kernel of the morphism

$$\mathbf{F}_p = \mathbf{Z}/p\mathbf{Z} \to A/\mathfrak{p}.$$

This morphism is injective, like any morphism of fields.

As $\overline{A}$ is of finite dimension over $\mathbf{F}_p$, the extension $k/\mathbf{F}_p$ is finite, and Galois, like any finite extension of finite fields. Just as A is generated by the polynomials in the z_i with coefficients in $\mathbf{Z}$, k is generated by the polynomials in the $x_i = (z_i \bmod \mathfrak{p})$ with coefficients in $\mathbf{F}_p$. In other words, we obtain:

Lemma 10.5.2 k *is the splitting field of* $\overline{P}$ *over* $\mathbf{F}_p$.

This, incidentally, proves again that k is of finite dimension over $\mathbf{F}_p$.

Now, the Galois group $G = \text{Gal}(K/\mathbf{Q})$ permutes the z_i and therefore A is stable under its action.

---

[2] Note that its existence is completely independent of the axiom of choice (use for example that A is of finite type over $\mathbf{Z}$).

Definition 10.5.3 The subgroup $D = D_\mathfrak{p}$ of G fixing $\mathfrak{p}$ is called the *decomposition group* of $\mathfrak{p}$.

As A is globally invariant under the action of D, this defines an action on the quotient $k = A/\mathfrak{p}$. We therefore have a group morphism $\phi : D \to \text{Gal}(k/\mathbf{F}_p)$.

Theorem 10.5.4 *The group morphism ϕ is surjective.*

Proof An element $\sigma_0 \in \text{Gal}(k/\mathbf{F}_p)$ is determined by the image $y = \sigma_0(x)$ of a generator $x \neq 0$ of the extension $k/\mathbf{F}_p$.

The ideals $g^{-1}(\mathfrak{p})$ are equal to $\mathfrak{p}$ if and only if $g \in D$. Moreover, the projection

$$A \xrightarrow{g} A \to A/\mathfrak{p}$$

is surjective because g is bijective and admits $g^{-1}(\mathfrak{p})$ as kernel. Thus, we have an isomorphism

$$A/g^{-1}(\mathfrak{p}) \xrightarrow{\sim} A/\mathfrak{p}$$

ensuring that $g^{-1}(\mathfrak{p})$ is maximal since the corresponding quotient is the field $A/\mathfrak{p}$.[3]

Let $\mathfrak{q}_1, \ldots, \mathfrak{q}_r$ be the (distinct) ideals of the form $g^{-1}(\mathfrak{p})$, $g \notin D$. As $\mathfrak{q}_0 = \mathfrak{p}, \mathfrak{q}_1, \ldots, \mathfrak{q}_r$ are pairwise distinct and maximal, we have $\mathfrak{q}_i + \mathfrak{q}_j = A$ if $i \neq j$. According to the Chinese Lemma (3.8.1), we can find $z \in A$ such that

$$z \equiv x \bmod \mathfrak{q}_0 \text{ and } z \equiv 0 \bmod \mathfrak{q}_i \text{ if } i > 0,$$

and therefore

$$z \equiv x \bmod \mathfrak{p} \text{ and } z \equiv 0 \bmod g^{-1}(\mathfrak{p}) \text{ if } g \notin D.$$

We then have $g(z) \in \mathfrak{p}$ if $g \notin D$. The polynomial

$$\prod_{g \in G}(X - g(z))$$

has integer coefficients, its coefficients being invariant under the action of G and integers over $\mathbf{Z}$. By construction, its image in $k[X] = A/\mathfrak{p}[X]$ is written

$$\prod_{g \in D}(X - \overline{g(z)}) \prod_{g \notin D} X$$

---

[3] In fact, it is not difficult to prove that the non-zero prime ideals of A are maximal, but we will not need this.

and annihilates $\bar{z} = x$. As x is non-zero, we deduce that the polynomial

$$\prod_{g \in D} (X - \overline{g(z)}) \in \mathbf{F}_p[X]$$

is divisible by the minimal polynomial

$$\prod_{\sigma \in \mathrm{Gal}(k/\mathbf{F}_p)} (X - \sigma(x))$$

of x over k, and that therefore there exists a $g \in D$ such that $\sigma_0(x) = \overline{g(z)}$, which is what we wanted.   □

Let us denote by $x_1, \ldots, x_n$ the reductions modulo $\mathfrak{p}$ of the roots $z_1, \ldots, z_n$ of P.

Theorem 10.5.5 *Suppose that* $\overline{P}$ *has simple roots (in* $\overline{\mathbf{F}}_p$*). Then,* ϕ *is an isomorphism of the subgroup* D *of* $\mathrm{Gal}(P, \mathbf{Q})$ *onto the group* $\mathrm{Gal}(\overline{P}, \mathbf{F}_p)$*. Moreover,* ϕ *is compatible with the embeddings of the Galois groups in the symmetric group* S_n *(cf. 10.5.1) defined by the numbering* $\{z_i\}_{1 \le i \le n}$ *of the roots of* P *and* $\{x_i\}_{1 \le i \le n}$ *of the roots of* $\overline{P}$*.*

Proof By hypothesis, the x_i are distinct. In other words, the mapping $z_i \mapsto x_i$ is bijective and induces an identification of the permutation groups

$$\mathrm{Bij}(\{z_i\}_{1 \le i \le n}) = \mathrm{Bij}(\{x_i\}_{1 \le i \le n}).$$

We have a diagram

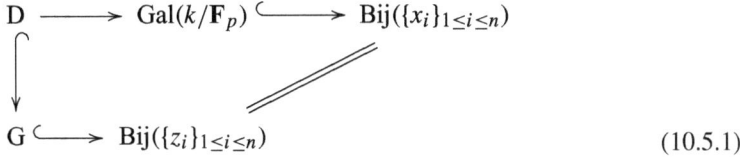

$$\hspace{14cm} (10.5.1)$$

which commutes, that is, the two maps composed from D are the same. This proves the injectivity of ϕ. Moreover, we already know that ϕ is surjective.   □

Remark 10.5.6 The proof of the lemma gives a little more. Assuming the hypotheses are satisfied, if we have a permutation of the roots of $\overline{P}$, there exists a permutation of the roots of P of the same type. In particular, if $\overline{P}$ is irreducible, there exists a cycle of length n in G according to Lemma 9.1.7.

Let us finally see that, despite appearances, the subgroup $D = D_{\mathfrak{p}}$ depends very little on $\mathfrak{p}$ but rather on p. We give ourselves two maximal ideals $\mathfrak{p}$, $\mathfrak{q}$ of A such that $A/\mathfrak{p} \xrightarrow{\sim} A/\mathfrak{q} \xrightarrow{\sim} \mathbf{F}_p$.

Proposition 10.5.7 *There exists a* $g \in G$ *such that* $\mathfrak{p} = g(\mathfrak{q})$. *We then have* $D_{\mathfrak{p}} = g D_{\mathfrak{q}} g^{-1}$.

Proof For the first part, imagine that we have $\mathfrak{p} \not\subset g(\mathfrak{q})$ for all $g \in G$. We then have $\mathfrak{p} + g(\mathfrak{q}) = A$ for all g because $\mathfrak{p}$ is maximal. The Chinese Lemma (3.8.1) allows us to construct $x \in A$ such that $x \equiv 1 \bmod g(\mathfrak{q})$ for all g and $x \equiv 0 \bmod \mathfrak{p}$. Then $N = \prod_{g \in G} g(x)$ is in

$$p\mathbf{Z} = \mathfrak{p} \cap \mathbf{Z} = \mathfrak{q} \cap \mathbf{Z}.$$

So $N \in \mathfrak{q}$ and therefore one of the factors $g(x)$ is in $\mathfrak{q}$. In other words $x \equiv 0 \bmod g^{-1}(\mathfrak{q})$, a contradiction. The second part follows easily. $\qquad\square$

In particular, $D_{\mathfrak{p}}$ and $D_{\mathfrak{q}}$ are isomorphic (via the inner automorphism $h \mapsto ghg^{-1}$), and even equal if G is abelian. Thus, the Frobenius morphism defines an element of G unique up to conjugation, and unique if G is abelian! This is the beginning of class field theory...

Theorem 10.5.5 often allows the Galois group of a polynomial to be calculated. We will see in particular in Sect. 10.6 an application to cyclotomy, and one can refer to Sect. 11.4 for an application to the Galois group of a polynomial with integer coefficients.

All this is not a coincidence: we will see in Sect. 10.7 why the method of reduction modulo p is so effective.

10.6   Cyclotomy and Reduction Modulo p

We will show here how the theory of reduction modulo p of Galois groups allows us to calculate the Galois group of the cyclotomic extension over $\mathbf{Q}$ (Chap. 8) and thus to reprove the irreducibility over $\mathbf{Q}$ of the cyclotomic polynomial.

Let $n \geq 1$ be an integer and $\zeta \in \mu_n(\mathbf{C})$ a primitive root of unity (for example $\zeta = \exp(\frac{2i\pi}{n})$). Let $K = \mathbf{Q}[\zeta]$ be the cyclotomic field: it is the splitting field of $X^n - 1$. We know (beginning of Chap. 8) that K is Galois over $\mathbf{Q}$ and there exists (Sect. 8.2) a canonical injective group morphism (the cyclotomic character on $\mathbf{Q}$)

$$\chi_{\mathbf{Q}} : G := \mathrm{Gal}(K/\mathbf{Q}) \hookrightarrow (\mathbf{Z}/n\mathbf{Z})^*$$

characterized by

$$\forall g \in G, \ g(\zeta) = \zeta^{\chi_{\mathbf{Q}}(g)}. \tag{10.6.1}$$

We will therefore show

Theorem 10.6.1 $\chi_{\mathbf{Q}}$ *is an isomorphism.*

Proof It suffices to show that $\chi_{\mathbf{Q}}$ is surjective. Recall that $(\mathbf{Z}/n\mathbf{Z})^*$ is the multiplicative group of classes $\bar{m}$ with $m \in \mathbf{Z}$ such that $(m, n) = 1$. By decomposing m into prime factors, we realize that the theorem comes down to proving that if p is prime and does not divide n, then $p \in \mathrm{Im}(\chi_{\mathbf{Q}})$.

Let p be a prime number not dividing n. Let A be the subring

$$A = \mathbf{Z}[\zeta]$$

of $\mathbf{C}$. Choose a maximal ideal $\mathfrak{p}$ of A such that $p \in \mathfrak{p}$ as in the previous section. Since n is coprime with p, the derivative polynomial of $X^n - 1$ is a non-zero multiple of X^n in $\mathbf{F}_p[X]$ so that $X^n - 1$ is separable. We can apply the theory of the reduction of Galois groups, and Theorem 10.5.5. We obtain that

$$A \to k(\mathfrak{p}) := A/\mathfrak{p}$$

induces an isomorphism $g \mapsto \bar{g}$ of the decomposition group

$$G \supset D_{\mathfrak{p}} \xrightarrow{\sim} \mathrm{Gal}(k(\mathfrak{p})/\mathbf{F}_p)$$

and that $k(\mathfrak{p})$ is the splitting field of $X^n - 1$. The isomorphism is characterized by the relation for all $a \in A$ and $g \in D_{\mathfrak{p}}$:

$$g(a \bmod \mathfrak{p}) = g(a) \bmod \mathfrak{p}. \tag{10.6.2}$$

There exists therefore a unique element $F_{\mathfrak{p}} \in D_p$ such that

$$\bar{F}_{\mathfrak{p}} = F, \tag{10.6.3}$$

where F is the Frobenius morphism of $k(\mathfrak{p})$. We have in $\mathbf{C}[X]$ the decomposition

$$X^n - 1 = \prod_{\xi \in \mu_n(\mathbf{C})} (X - \xi),$$

which is an equality in A[X]. By reducing it modulo p, we obtain the decomposition

$$X^n - 1 = \prod_{\xi \in \mu_n(\mathbf{C})} (X - \bar{\xi}),$$

which is an equality in $k(\mathfrak{p})[X]$. Since $X^n - 1$ is separable over $\mathbf{F}_p$, we deduce that the reduction morphism modulo $\mathfrak{p}$

$$\begin{cases} \mu_n(\mathbf{C}) \to \mu_n(k(\mathfrak{p})) \\ \xi \quad \mapsto \bar{\xi} \end{cases}$$

is injective and therefore bijective for reasons of cardinality. The image of a generator is therefore a generator so that $\bar{\zeta}$ is a primitive n^{th} root of unity in $k(\mathfrak{p})$. The cyclotomic theory on $\mathbf{F}_p$ (Chap. 8) then assures that there exists a canonical injective morphism (the cyclotomic character on $\mathbf{F}_p$)

$$\chi_p : \mathbf{D}_\mathfrak{p} = \mathrm{Gal}(k(\mathfrak{p})/\mathbf{F}_p) \hookrightarrow (\mathbf{Z}/n\mathbf{Z})^*$$

characterized by

$$\forall \gamma \in \mathbf{D}_\mathfrak{p}, \ \gamma(\bar{\zeta}) = \bar{\zeta}^{\chi_p(\zeta)}. \tag{10.6.4}$$

By comparing formulas (10.6.1) and (10.6.4) with (10.6.3) and (10.6.2), we obtain the compatibility formula

$$\forall g \in \mathbf{D}_\mathfrak{p}, \ \chi_\mathbf{Q}(g) = \chi_p(\bar{g}).$$

As $F(\bar{a}) = a^p$ for all $a \in A$, we have by definition

$$\chi_p(F) = p$$

so that $\chi_\mathbf{Q}(F_\mathfrak{p}) = p$.   □

10.7   The Chebotarev Theorem

Let us conclude this journey with a paragraph without proof, explaining why the method of reduction modulo p is so effective.

Let P be a separable polynomial with integer coefficients. Let K be the splitting field of P. It is a Galois extension of $\mathbf{Q}$. If p is a sufficiently large prime number (neither dividing the leading coefficient of P nor its discriminant), let's say $p > n(P)$, the reduction $\bar{P}$ modulo p of P has simple roots. The decomposition group $\mathbf{D}_\mathfrak{p} = \mathrm{Gal}(k/\mathbf{F}_p)$ is therefore cyclic, generated by the Frobenius morphism, well defined up to conjugation. Let C_p be the set of elements of $\mathrm{Gal}(P, \mathbf{Q})$ conjugate to such an element (which depends only on p and not on $\mathfrak{p}$). Let C be the conjugacy class of an element of G. We may wonder if C comes from the characteristic p, in other words, if $C = C_p$. Chebotarev (Fig. 10.2) proved that this is true with "probability" card C/ card G, in the following sense.

Fig. 10.2 Nikolai
Gregorievich Chebotarev
(1894–1947). Author
unknown. Source: Wikimedia
Commons

Theorem 10.7.1 (Chebotarev) *The limit of the sequence*

$$n \mapsto \frac{\mathrm{card}\{p \text{ primes such that } C = C_p \text{ and } n(\mathrm{P}) < p \le n\}}{\mathrm{card}\{p \text{ primes such that } p \le n\}}$$

exists and is equal to $\mathrm{card}\,C / \mathrm{card}\,G$.

The proof uses more refined techniques than those developed here.

For example, if $C = \{1\}$, this "probability"[4] is $1/\mathrm{card}\,G$. It is easy to see that the condition that C_p is trivial means that $\overline{\mathrm{P}}$ is split. In particular, this theorem says that there are infinitely many p such that $\overline{\mathrm{P}}$ is split over $\mathbf{F}_p$, which can be demonstrated in an elementary, but clever way. These are "the bad prime numbers p" from the point of view of the calculation of the Galois group.

Exercise 10.7.2 Show that if an integer is a square modulo p for all sufficiently large prime numbers p, then it is a square. Show a similar result for l-th powers, where l is a prime number (difficult).

---

[4] More precisely, it is a density (see [Ser87]).

Chapter 11
Complements

11.1 Zorn's Lemma and Applications

Let E be a (partially) ordered set. We might think, for example, of the set of subsets of a given set ordered by inclusion. But there are many other examples.

Definition 11.1.1 We say that E is *inductive* if every non-empty totally ordered subset of E has an upper bound in E.

Example 11.1.2 $\mathbf{R}$ equipped with the usual order relation is not inductive. Similarly, the set of intervals $[0, x[$, $x \in \mathbf{R}$, ordered by inclusion is not inductive. On the other hand, the set of subsets of a set ordered by inclusion is inductive.

Lemma 11.1.3 (Zorn's Lemma) *Every non-empty inductive set has a maximal element.*

This lemma due to Max Zorn (Fig. 11.1) can be seen as an axiom of set theory, in fact it equivalent to the axiom of choice: if (E_i) is a non-empty family of sets, then $\prod E_i$ is non-empty. We will consider it as such.

Corollary 11.1.4 *Every non-zero ring has a maximal ideal. More generally, every proper ideal of a ring is contained in a maximal ideal.*

© The Author(s), under exclusive license to Springer Nature Switzerland AG 2024
D. Hernandez, Y. Laszlo, *Introduction to Galois Theory*, Springer Undergraduate
Mathematics Series, https://doi.org/10.1007/978-3-031-66182-2_11

Fig. 11.1 Max Zorn (1906–1993). Author: Paul R. Halmos. Source: Paul R. Halmos photograph collection, The Dolph Briscoe Center for American History, The University of Texas at Austin

Proof Let E be the family of proper ideals of A. As A is non-zero, {0} is in E, which is thus non-empty. Obviously, E is inductive: the union of a totally ordered family of proper ideals is still a proper ideal, which is an upper bound. Zorn's Lemma finishes the job. Considering A/I, we find that every proper ideal I is contained in a maximal ideal (simply because the ideals (resp. maximal ideals) of A/I identify with the ideals (resp. maximal ideals) of A containing I according to Lemma 3.4.6).

$\square$

11.2 Galois Group of Composite Extensions

In the study of cyclotomic extensions, we encountered the problem of studying a composite extension KL/k in terms of K/k and L/k (when these last two are cyclotomic extensions over $\mathbf{Q}$). We can do this in general (we could also deduce in this way the calculation of $\mathbf{Q}[\zeta_n] \cap \mathbf{Q}[\zeta_m]$ carried out in Proposition 8.4.1).

We assume that all the extensions considered are contained in an algebraically closed extension Ω of a field k.

If x_i is a family of elements of Ω, the intersection of the subfields of Ω containing the x_i is the smallest subfield of Ω containing the x_i. If K, L are two extensions of k, the smallest field containing K, L is denoted KL and is called the composite extension of K and L.

Lemma 11.2.1 *Let* K *be a finite extension of* k. *There exists a unique subfield of* Ω *containing* K *which is Galois over* k: *we call it the Galois closure of* K/k.

Proof The intersection of two Galois extensions of k is clearly still Galois. Uniqueness follows from this. For existence, let us choose a primitive element x

of K/k (recall that k is perfect). The splitting field of the minimal polynomial of x over k is the sought extension.   □

Theorem 11.2.2 *Let* K, L *be two finite extensions of* k *and suppose* K/k *is Galois.*

(i) *Then,* KL/L *is Galois and the restriction morphism*

$$r : \mathrm{Gal}(\mathrm{KL}/\mathrm{L}) \to \mathrm{Gal}(\mathrm{K}/\mathrm{K} \cap \mathrm{L})$$

   is an isomorphism.
(ii) *If in addition* L/k *is Galois, then* KL/k *and* K ∩ L/k *are also Galois.*

Proof

(i) Let x be a primitive element of K/k with minimal polynomial $P \in k[X]$. Clearly, $KL = L[x]$ and the minimal polynomial Q of x over L divides P so that its roots are in K (like those of P) and *a fortiori* in KL. As P has simple roots, this proves that KL/L is Galois. More specifically, we have $Q = \prod(X - x_i)$ where $x_i \in K$ are certain conjugates of x. Therefore Q is also in $K[X]$, which proves that we have $Q \in (K \cap L)[X]$.

   Let us prove the surjectivity of r. Let $g \in \mathrm{Gal}(K/K \cap L)$. We have $g(Q(x)) = Q(g(x))$ because Q has coefficients in $K \cap L$. By the universal property of the quotient, there exists a unique L-endomorphism of $KL = L[X]/Q$ that sends $x = (X \bmod Q)$ to $g(x)$: this is the sought pre-image.

   We now prove the injectivity of r. Let $g \in \mathrm{Gal}(KL/L)$ be in the kernel, that is, trivial on K. As g is trivial on L and as K, L generate KL, it is trivial on KL, which is what we wanted.

(ii) Suppose in addition L is Galois. Then, L is the splitting field of a separable polynomial $P_1 \in k[X]$ and KL is the splitting field of the separable polynomial $LCM(P, P_1)$, proving that KL/k is Galois. For the last point, let $\sigma \in \mathrm{Hom}_k(K \cap L, \Omega)$, which we extend to the whole KL. As K, L are Galois over k, we have $\sigma(K) \subset K$ and $\sigma(L) \subset L$ and therefore $\sigma(K \cap L) \subset K \cap L$, from which we get the equality (since $K \cap L$ is finite-dimensional over k).   □

Item (i) of the theorem can be represented graphically:

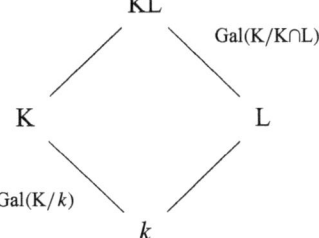

Corollary 11.2.3 *Under the assumptions of the theorem, we have*

$$[KL : L] = [K : K \cap L] \text{ and } [KL : k] = [K : k][L : k]/[K \cap L : k].$$

Consequently, $[KL : k] = [K : k][L : k]$ if and only if $k = K \cap L$.

Proof The first equation follows from the previous item ii). For the second, we then write

$$[KL : k] = [KL : L][L : k] = [K : K \cap L][L : k] = ([K : k]/[K \cap L : k])[L : k].$$

The third follows from this.   □

Let us denote by i : $\mathrm{Gal}(KL/k) \rightarrow \mathrm{Gal}(K/k) \times \mathrm{Gal}(L/k)$ the restriction morphism, which is clearly injective. The restriction morphisms

$$\mathrm{Gal}(K/k) \rightarrow \mathrm{Gal}(K \cap L/k) \text{ and } \mathrm{Gal}(L/k) \rightarrow \mathrm{Gal}(K \cap L/k)$$

define a morphism

$$(j_1, j_2) : \mathrm{Gal}(K/k) \times \mathrm{Gal}(L/k) \rightarrow \mathrm{Gal}(L \cap K/k) \times \mathrm{Gal}(L \cap K/k).$$

Of course, the first component (resp. second component) of i composed with j_1 (resp. j_2) coincides with the natural restriction morphism

$$\mathrm{Gal}(KL/k) \rightarrow \mathrm{Gal}(K \cap L/k).$$

Proposition 11.2.4 *Suppose K/k and L/k are Galois. The morphism (injective) i induces an isomorphism of $\mathrm{Gal}(KL/k)$ onto the subgroup*

$$\mathrm{Gal}(K/k) \times_{\mathrm{Gal}(K \cap L/k)} \mathrm{Gal}(L/k)$$

of $\mathrm{Gal}(K/k) \times \mathrm{Gal}(L/k)$ consisting of pairs (u, v) such that $j_1(u) = j_2(v)$ (this subgroup is called the amalgamated product).

Proof It suffices to show that the cardinalities of the two groups in question are equal. We denote by $G_L, G_K, \ldots$ the Galois groups over k. We have an exact sequence

$$1 \rightarrow N \rightarrow G_K \times G_L \overset{(j_1, j_2)}{\rightarrow} G_{K \cap L} \times G_{K \cap L} \rightarrow 1$$

(note that j_1, and *a fortiori* (j_1, j_2), is surjective). By construction, our amalgamated product G is the inverse image by (j_1, j_2) of the diagonal subgroup

$$G_{K \cap L} = \{(g, g), g \in G_{K \cap L}\} \subset G_{K \cap L} \times G_{K \cap L}.$$

We therefore have an exact sequence

$$1 \rightarrow N \rightarrow G \rightarrow G_{K \cap L} \rightarrow 1.$$

Comparing cardinalities, we deduce

$$\text{card}\, G = [K : k][L : k]/[K \cap L : k]$$

and the result follows thanks to Corollary 11.2.3. □

This statement is very pleasant when in addition $K \cap L = k$, so that the amalgamated product is none other than the usual product.

The theorem can be graphically represented as follows

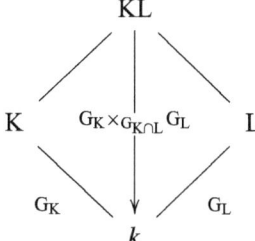

which can be completed according to the above as

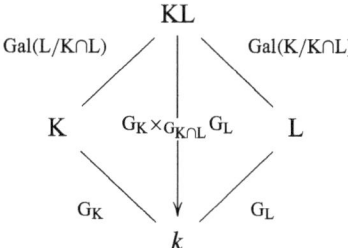

11.3 Transcendence of e and π

The methods of proving the transcendence of e and π are analogous.

Let P be a polynomial of degree m with real coefficients and $\tilde{P}$ the polynomial derived from P by replacing each coefficient with its absolute value. We then set

$$I(t) = \int_0^t e^{t-u} P(u)\, du.$$

We have (integration by parts)

$$I(t) = e^t \sum_{j=0}^{m} P^{(j)}(0) - \sum_{j=0}^{m} P^{(j)}(|t|) \tag{11.3.1}$$

and

$$|I(t)| \leq |t| e^{|t|} \tilde{P}(|t|). \tag{11.3.2}$$

11.3.1 Transcendence of e

Suppose that e is algebraic. We then have

$$\sum_{i=0}^{n} a_i e^i = 0 \tag{11.3.3}$$

with $n, a_0 > 0$ and the a_i integers. Let

$$J = \sum_{k=0}^{n} a_k I(k).$$

With the previous notations, formulas (11.3.1) and (11.3.3) immediately give

$$J = - \sum_{j=0}^{m} \sum_{k=0}^{n} a_k P^{(j)}(k)$$

which already ensures that J is an integer.
  We choose $p > n a_0$ prime and we define

$$P(X) = X^{p-1}(X-1)^p \cdots (X-n)^p$$

and so $m = (n+1)p - 1$.
  By construction, we have
$P^{(j)}(k) = 0$ if $j < p$ and $k > 0$,
$P^{(j)}(0) = 0$ if $j < p - 1$.
  Thus

$$J = -a_0 P^{(p-1)}(0) - \sum_{j=p}^{m} \sum_{k=p}^{n} a_k P^{(j)}(k).$$

However, $j!|P^{(j)}(k)$ for all integers j, k (Taylor's formula for example) so that

$$(p-1)!|J \text{ and } J \equiv -a_0 P^{(p)}(0) \text{ mod } (p!).$$

A direct calculation also gives

$$a_0 P^{(p-1)}(0)/(p-1)! = \pm a_0(n!)^p.$$

As $p > na_0$, it does not divide the integer $a_0 P^{(p-1)}(0)/(p-1)!$, which is therefore non-zero, hence ≥ 1. We therefore have the inequality

$$|J| \geq (p-1)!$$

A direct calculation also shows that we have

$$\tilde{P}(k) \leq (2n)^m$$

for $0 \leq k \leq n$. From this and (11.3.2), we deduce the existence of c depending only on the a_i and n (and not on p) such that

$$|J| \leq c^p$$

for all sufficiently large prime p, which contradicts $|J| \geq (p-1)!$.

11.3.2 Transcendence of π

Suppose that π (and hence also $I\pi$) is algebraic over $\mathbf{Q}$. Let $\alpha_1, \ldots, \alpha_d$ be the conjugates of $I\pi$ and G the Galois group over $\mathbf{Q}$ of the field they generate. If $N > 0$ is a common denominator of the coefficients of the minimal polynomial of $I\pi$, the $N\alpha_j$ are algebraic integers. For all $\epsilon = (\epsilon_i) \in \{0, 1\}^d$, let

$$\alpha_\epsilon = \sum_j \epsilon_j \alpha_j.$$

We have

$$0 = \prod_j (1 + \exp(\alpha_j)) = \sum_\epsilon \exp(\alpha_\epsilon) = q + \sum_{\epsilon \in A} \exp(\alpha_\epsilon) \qquad (11.3.4)$$

where $A = \{\epsilon | \alpha_\epsilon \neq 0\}$ and $q = 2^d - \text{card}(A) = 2^d - n > 0$. Let $a_1, \ldots, a_n$ be the n elements of A.

Lemma 11.3.1 *Let* $S \in \mathbf{Z}[X_1, \ldots, X_n]$ *be a symmetric polynomial in the* X_i. *Then,* $s = S(Na_1, \ldots, Na_n) \in \mathbf{Z}$.

Proof As G permutes the α_j, it also permutes the elements of A. This implies that s is fixed by G therefore is rational (Proposition 6.4.1). But S is also an algebraic integer, therefore it is in $\mathbf{Z}$ (Sect. 10.3). □

We proceed as above with

$$P(X) = N^{np}X^{p-1}(x - a_1)^p \cdots (X - a_n)^p$$

where p is a prime number intended to become large and $m = (n+1)p+1$. Thanks to (11.3.1) and (11.3.4), we have

$$J := I(a_1) + \cdots + I(a_n) = -q \sum_{j=0}^{m} P^{(j)}(0) - \sum_{j=0}^{m}\sum_{k=1}^{n} P^{(j)}(a_k).$$

We proceed exactly as above by noting that 0 and β_i are roots of order greater than or equal to $p - 1$, which, thanks to Lemma 11.3.1, ensures

$$(p - 1)! | J.$$

As above, we must show that p does not divide the integer (Lemma 11.3.1)

$$qP^{(p-1)}(0)/(p - 1)! = \pm qN^{np-n}(Na_1 \cdots Na_n)$$

which is the case if p is large enough. Thus, $|J/(p - 1)!|$ is a non-zero integer and therefore ≥ 1. As above, we obtain thanks to (11.3.2) the existence of a constant c independent of p such that $|J| \leq c^p$, which contradicts $|J| \geq (p - 1)!$ for p large enough.

11.4 The Galois Group Over Q of a Polynomial with Integer Coefficients

The aim of this section is to study the Galois group over $\mathbf{Q}$ of polynomials with integer coefficients.

First, let us prove the following lemma, a generalization of Lemma 9.1.7.

Lemma 11.4.1 *Let* $P_1, \ldots, P_r \in \mathbf{F}_p[X]$ *be of respective degrees* $d_1, \ldots, d_r > 0$, *irreducible and pairwise coprime. Let* K *be the splitting field of* $\prod_{1 \leq i \leq r} P_i$, G = Gal$(K, \mathbf{F}_p)$ *and* $F \in G$ *the Frobenius morphism of* K. *Then* G *is canonically a* *subgroup of the product group* $S_{d_1} \times \cdots \times S_{d_r}$, F *is a product of disjoint cycles of length* d_i *and*

$$\mathrm{card}(G) = \mathrm{LCM}_{1 \leq i \leq r}(d_i).$$

Proof Since P_i is irreducible and $\mathbf{F}_p$ is perfect, P_i is separable. For $1 \leq i \leq r$, let $(z_{i,j})_{1 \leq j \leq d_i}$ be the roots of P_i in $\overline{\mathbf{F}}_p$. As the polynomials are pairwise coprime, the $z_{i,j}$ are all distinct. Since G acts on each set $(z_{i,j})_{1 \leq j \leq d_i}$, we can embed G in the product group $S_{d_1} \times \cdots \times S_{d_r}$ where we identify $S_{d_i} = \mathrm{Bij}(z_{i,1}, \ldots, z_{i,d_i})$. As P_i is an irreducible polynomial, P_i is the minimal polynomial of $z_{i,1}$ over $\mathbf{F}_p$ and its roots are the conjugates of $z_{i,1}$. The field $\mathbf{F}_p$ being finite, G is generated by the Frobenius morphism F (Theorem 5.2.3). The conjugates of $z_{i,1}$ are therefore exactly the $F^n(z_{i,1})$, $n = 0, \ldots, d_i - 1$. Let γ_i be the cycle

$$(z_{i,1}, F(z_{i,1}), \ldots, F^{d_i - 1}(z_{i,1}))$$

in $S_{d_i} = \mathrm{Bij}(z_{i,1}, \ldots, z_{i,d_i})$. By construction, we have $F = \gamma_1 \times \cdots \times \gamma_r$ viewed in $S_{d_1} \times \cdots \times S_{d_r}$. As the γ_i pairwise commute (they have disjoint supports), we have $F^n = \prod_{1 \leq i \leq r} \gamma_i^n$ for all $n \geq 0$. This ensures that the order of F is the LCM of the d_i. As F generates G, we deduce that $\mathrm{card}(G) = \mathrm{LCM}_{1 \leq j \leq r}(d_i)$. $\square$

Let now P be a polynomial with integer coefficients of degree $n \geq 3$. We assume that there exist 3 prime numbers p_0, p_1, p_2 such that for $0 \leq i \leq 2$, the reduction of P modulo p_i has a unique irreducible factor of degree less than or equal to d_i, with $d_0 = n$, $d_1 = n - 1$ and $d_2 = 2$.

Proposition 11.4.2 *The Galois group of* P *over* **Q** *is* S_n.

Proof Theorem 10.5.5 and Lemma 11.4.1 ensure that the Galois group $G \subset S_n$ contains an n-cycle, an $n - 1$ cycle and a transposition. The result then follows by Exercise 2.6.5. $\square$

In fact, by (barely) refining the previous proof, we can show, in a suitable sense, that the probability[1] that a polynomial of degree n with integer coefficients has S_n for its Galois group over **Q** is 1 (see [Bou23, Exercise V.12.13])!

11.5 Symmetric Polynomials

The symmetric group S_n acts by permutation of the indices on the ring $\mathbf{C}[X_1, \ldots, X_n]$ of polynomials with n indeterminates $X_1, \ldots, X_n$. A symmetric polynomial is a polynomial in $\mathbf{C}[X_1, \ldots, X_n]$ that is fixed under the action of S_n.

---

[1] We should rather speak of the *density*.

The ring of symmetric polynomials is thus the set of fixed points of the action of S_n, that is $(\mathbf{C}[X_1, \ldots, X_n])^{S_n}$ (it is immediately verified that this is indeed a subring).

For example, the elementary symmetric polynomials σ_i defined by the formula (9.4.1) are indeed in $(\mathbf{C}[X_1, \ldots, X_n])^{S_n}$, because the product $\prod_{i=1}^{n}(X - X_i)$ is clearly fixed by the action of S_n on its coefficients. Note that $\sigma_n = X_1 X_2 \cdots X_n$ and $\sigma_1 = X_1 + X_2 + \cdots + X_n$.

Theorem 11.5.1 (Theorem of Symmetric Polynomials) *The ring of symmetric polynomials is generated by the elementary symmetric polynomials:*

$$(\mathbf{C}[X_1, \ldots, X_n])^{S_n} = \mathbf{C}[\sigma_1, \ldots, \sigma_n].$$

Proof The degree of a symmetric polynomial is the maximum degree d of the monomials $X_1^{d_1} \cdots X_n^{d_n}$ that compose it, where $d = d_1 + \cdots + d_n$. We can then prove the result by induction on n and on d as in [Lan02, IV.6]. The result for $n = 1$ is trivial. Let $P(X_1, \ldots, X_n)$ be symmetric of degree d. Then $P(X_1, \ldots, X_{n-1}, 0)$ is symmetric in $X_1, \ldots, X_{n-1}$. According to the induction hypothesis on n, we have therefore $P(X_1, \ldots, X_{n-1}, 0) = Q(\sigma_1', \ldots, \sigma_{n-1}')$ with Q a polynomial of $n - 1$ variables and $\sigma_1', \ldots, \sigma_{n-1}'$ the elementary symmetric polynomials in $X_1, \ldots, X_{n-1}$. Note that σ_i' is obtained from σ_i by setting $X_n = 0$. Now consider

$$\tilde{P}(X_1, \ldots, X_n) = P(X_1, \ldots, X_n) - Q(\sigma_1, \ldots, \sigma_{n-1}).$$

This is a symmetric polynomial such that $\tilde{P}(X_1, \ldots, X_{n-1}, 0) = 0$. It is therefore divisible by X_n, and therefore by $\sigma_n = X_1 \cdots X_n$ by symmetry, that is $\tilde{P}(X_1, \ldots, X_n) = \sigma_n \tilde{Q}(X_1, \ldots, X_n)$ with $\tilde{Q}$ symmetric of degree strictly less than that of $\tilde{P}$. The induction hypothesis on the degree d completes the proof.   □

A fraction $R = P/Q$ is invariant under the action of S_n if and only if it can be written as a quotient of two invariant polynomials (write $R = (P \prod_{\sigma \neq \mathrm{Id}} \sigma.Q)/(\prod \sigma.Q)$). In terms of the field of fractions, we therefore have

$$(\mathrm{Frac}(\mathbf{C}[X_1, \ldots, X_n]))^{S_n} = \mathrm{Frac}((\mathbf{C}[X_1, \ldots, X_n])^{S_n}) = \mathrm{Frac}(\mathbf{C}[\sigma_1, \ldots, \sigma_n]).$$

Fig. 11.2 Andrew Wiles
(1953–). Author: C. J.
Mozzochi. Source:
Wikimedia Commons. © C. J.
Mozzochi, Princeton N.J

11.6   Some Words on Inverse Galois Theory

Let $\overline{\mathbf{Q}}$ be the algebraic closure of $\mathbf{Q}$ in $\mathbf{C}$. The properties of the *absolute* Galois group

$$G = \mathrm{Gal}(\overline{\mathbf{Q}}/\mathbf{Q})$$

have profound arithmetic consequences.

For example, it was by studying the subtle properties of certain linear representations[2] of G in a $\mathbf{Q}_p$-vector space V of dimension 2 that Wiles (Fig. 11.2) was able to prove, among other things, Fermat's Last Theorem, an enigma that was more than 350 years old:

$$\text{if } x^n + y^n = z^n \text{ with } n \geq 3 \text{ and } x, y, z \in \mathbf{Z}, \text{ then } xyz = 0.$$

It is not possible to give an overview of the proof here, which far exceeds the level of this course. To try to understand G, we can already ask ourselves what its finite quotients are. This is what is called inverse Galois theory. It is an active research topic. Let us give an overview of it to finish this course. We will use without further caution the following principle, immediately deduced from Remark 6.1.10 :

Proposition 11.6.1 *Every quotient of the Galois group of a Galois extension of* $\mathbf{Q}$ *is isomorphic to a quotient of* G.

---

[2] That is, group morphisms from G to GL(V).

11.6.1   The Finite Abelian Case

We will prove the following statement.

Proposition 11.6.2 *Every finite abelian group is isomorphic to a quotient of* G.

Proof The Galois group G_n of the cyclotomic extension is $(\mathbf{Z}/n\mathbf{Z})^*$. To show that every finite abelian group is a quotient of G, it suffices to prove that every abelian group is a quotient of $H = (\mathbf{Z}/n\mathbf{Z})^*$ for suitable n.

Suppose that $n = p_1 \cdots p_m$ is a product of distinct prime numbers. According to the Chinese Lemma and Proposition 5.2.1, H is isomorphic to the product group

$$(\mathbf{Z}/(p_1 - 1)\mathbf{Z}) \times \cdots \times (\mathbf{Z}/(p_m - 1)\mathbf{Z}).$$

If N_i is an integer dividing $p_i - 1$, the reduction morphism modN$_i$ realizes $\mathbf{Z}/N_i\mathbf{Z}$ as a quotient of $\mathbf{Z}/(p_i - 1)\mathbf{Z}$. Thus, if $N_1, \ldots, N_m$ are integers dividing respectively $p_1 - 1, \ldots, p_m - 1$, we deduce that $\Pi = \prod_i \mathbf{Z}/N_i\mathbf{Z}$ is a quotient of H.

Now, let us give ourselves $N_1, \ldots, N_m$ strictly positive integers. Dirichlet's arithmetic progression theorem (cf. Exercise 8.3.13) ensures that we can find arbitrarily large prime numbers in each of the arithmetic progressions $1 + \lambda N_i$, $\lambda \in \mathbf{N}$. We can therefore choose $p_1, \ldots, p_m$ distinct such that $N_i | p_i - 1$ for all i, ensuring that Π is indeed a quotient of H, hence of G.

Now, every finite abelian group is a product of cyclic groups according to Proposition 2.3.1.   □

11.6.2   The First Non-abelian Non-trivial Case

The only non-abelian group of order ≤ 7 is $S_3 = D_6$, which is the Galois group of $X^3 - 2$ (Exercise 9.2.2), so it is a quotient of G (see Example 2.4.4 for the definition of D_n). There are 5 groups of order 8 up to isomorphism. Three are abelian, namely

$$\mathbf{Z}/8\mathbf{Z}, \mathbf{Z}/2\mathbf{Z} \times \mathbf{Z}/4\mathbf{Z}, \mathbf{Z}/2\mathbf{Z} \times \mathbf{Z}/2\mathbf{Z} \times \mathbf{Z}/2\mathbf{Z},$$

and two are not, namely

$$D_8, H_8.$$

The group H_8, known as the quaternion group, is the group with eight elements

$$1, i, j, k, t, ti, tj, tk,$$

where t is central[3] and

$$t^2 = 1, \text{ and } i^2 = j^2 = k^2 = ijk = t.$$

We have seen that D_8 is the Galois group of $X^4 - 2$, so that D_8 is a quotient of G. To distinguish them, it suffices to note that D_8 has 5 elements of order 2 while H_8 has only one, which generates its center (the subgroup of central elements).

We then have the following exercise.

Exercise 11.6.3 We aim to show that the field extension

$$\mathbf{Q}(\sqrt{(2 + \sqrt{2})(3 + \sqrt{6})})/\mathbf{Q}$$

is Galois with the Galois group being the group H_8.

(1) Let

$$a = (2 + \sqrt{2})(3 + \sqrt{6}),$$

and let $K = \mathbf{Q}(a)$: explain why the extension $\mathbf{Q}$ is Galois with the Galois group being the product of two cyclic groups of order 2. We will denote by

$$si, sj, sk \in \text{Gal}(K/\mathbf{Q})$$

the three non-trivial elements.
(2) Show, for each $\sigma = \sigma_i, \sigma_j, \sigma_k$, that $\sigma(a)/a$ is the square of an element of K that will be specified.
(3) Let $d = \sqrt{a}$ and $L = \mathbf{Q}(d)$. Show that $d \notin K$ (the previous question can be used). What is the Galois group of L/K? We denote its generator by τ, which we will consider as an element of $\text{Gal}(L/\mathbf{Q})$ (of which $\text{Gal}(L/K)$ is a subgroup).
(4) Define automorphisms $\tilde{\sigma}_i$ and $\tilde{\sigma}_j$ of L that extend σ_i and σ_j respectively. We will set $\tilde{\sigma}_k = \tilde{\sigma}_i \tilde{\sigma}_j$.
(5) Calculate the law of the group and conclude the group is H_8.

11.6.3 The Finite Reductive Case

The fundamental exact sequence (6.5.1) of Galois theory might suggest that we can deduce from the abelian case that every solvable group is isomorphic to the Galois group of an extension of $\mathbf{Q}$. Shafarevich (Fig. 11.3) proved that this is true, but the proof is much more complicated than just a simple reduction to the abelian case.

---

[3] t commutes with all elements of the group.

Fig. 11.3 Igor Rostilavovich Shafarevich (1923–2017). Author: Konrad Jacobs. Source: Konrad Jacobs and the Mathematisches Forschungsinstitut Oberwolfach. © MFO

Fig. 11.4 Walter Feit (1930–2004). Author unknown. Source: Yale University, Department of Mathematics

Fig. 11.5 John Griggs Thompson (1932–). Author: Renate Schmid. Source: Mathematisches Forschungsinstitut Oberwolfach. © MFO

Theorem 11.6.4 (Shafarevich) *Every finite solvable group is a quotient of* G.

There are many non-abelian solvable groups, such as the dihedral groups (Example 2.4.4). The mathematicians Feit (Fig. 11.4) and Thompson (Fig. 11.5) have shown the following (very difficult) result, which provides a simple way to identify if certain groups are solvable...

Experts conjecture that in fact every finite group is a quotient of G.

Theorem 11.6.5 (Feit–Thompson) *Every group of odd order is solvable.*

11.6.4   Some Quotients of G

Among the solvable groups we find A_n, S_n for $n \leq 4$. We have seen that S_n is a quotient of G (Proposition 11.4.2). The question of whether A_n is a quotient of G is very difficult and was solved by Hilbert, who introduced a very powerful method for constructing quotients of G from geometry.

Theorem 11.6.6 (Hilbert) *The alternating groups are quotients of* G.

It is known that A_n $(n \geq 5)$ is simple, in other words, it does not have a non-trivial quotient. The finite simple groups have been classified. In addition to the alternating groups, there is an infinite list of matrix groups with coefficients in finite fields, such as the groups

$$PSL_n(\mathbf{F}_q) = SL_n(\mathbf{F}_q)/\mu_n(\mathbf{F}_q)$$

and a finite list of 26 so-called sporadic groups. Among these, the largest of them, discovered in 1973, is called the "monster" and has cardinality

808 017 424 794 512 875 886 459 904 961 710 757 005 754 368 000 000 000.

All sporadic groups are quotients of G with one exception: as of January 2009, it is not known whether the Mathieu group[4] M_{23}, although relatively small, is a quotient of G, even though its cardinality "is only" 10 200 960 (compare to that of the monster!). In fact, it is not even known in general, far from it, whether the groups $GL_n(\mathbf{F}_q)$, or their avatars $PSL_n(\mathbf{F}_q)$, are quotients of G, even though many cases are known (see [Vol92] for results in this case, and J.-P. Serre (Fig. 11.6) [Ser87] for the general problem). It seems that the experts on the subject do not know, for example, whether the groups

$$PSL_2(\mathbf{F}_{5^3}) \text{ or } GL_4(\mathbf{F}_{2^2})$$

are quotients of G.

---

[4] named after Émile Léonard Mathieu (1835–1890).

Fig. 11.6 Jean-Pierre Serre (1926–). Author: Patrick Imbert. Source: Archives, Collège de France.
© Patrick Imbert/Collège de France

Chapter 12
Review Exercises

Here we gather exercises which are mostly taken from the exams given at the "École Polytechnique" at the end of the course from which this book is derived. Answers and solutions for these exercises are given in the following chapter.

Exercise 12.1 Let K be the field $\mathbf{Q}[\sqrt{2}]$.

(1) Show that if $x \in$ K is a square in K, then $N_{K/\mathbf{Q}}(x)$ (the determinant of the $\mathbf{Q}$-linear map $x \mathrm{Id}_K$) is a square in $\mathbf{Q}$. Deduce that $4 + 2\sqrt{2}$ is not a square in K.

Let L be the field $\mathbf{Q}[\sqrt{4 + 2\sqrt{2}}]$.

(2) Calculate $[L : \mathbf{Q}]$. What is the minimal polynomial of $\sqrt{4 + 2\sqrt{2}}$ over $\mathbf{Q}$? Over K?
(3) Show that $\sqrt{4 - 2\sqrt{2}} \in$ L. Deduce that $L/\mathbf{Q}$ is Galois.
(4) Show the existence of a unique $g \in \mathrm{Gal}(L/\mathbf{Q})$ such that

$$g(\sqrt{4 + 2\sqrt{2}}) = \sqrt{4 - 2\sqrt{2}}.$$

What is the order of g?
(5) What are the subfields of L. How many are there?

Exercise 12.2 Let $d_i, i = 1, \ldots, n$ be rational numbers. We assume that for any non-empty subset J of $\{1, \ldots, n\}$, the product $\prod_{j \in J} d_j$ is not a square in $\mathbf{Q}$. Let $K_i = \mathbf{Q}[\sqrt{d_1}, \ldots, \sqrt{d_i}]$. We want to show that we have $[K_n : \mathbf{Q}] = 2^n$ by induction. We have $K_0 = \mathbf{Q}$.

(1) Provide for all n an example of such a family $d_i, i = 1, \ldots, n$.
(2) Show that $K_n/\mathbf{Q}$ is Galois.

Assume $n \geq 1$ and the result has been proven for any family of such d_i of cardinality $\leq n$. Consider such a family of cardinality $n + 1$.

© The Author(s), under exclusive license to Springer Nature Switzerland AG 2024
D. Hernandez, Y. Laszlo, *Introduction to Galois Theory*, Springer Undergraduate
Mathematics Series, https://doi.org/10.1007/978-3-031-66182-2_12

(3) Show that $\mathrm{Gal}(K_n/K_{n-1})$ is isomorphic to $\mathbf{Z}/2\mathbf{Z}$.

Let σ be non-zero in $\mathrm{Gal}(K_n/K_{n-1})$.

(4) Assume $\sqrt{d_{n+1}} \in K_n$. Show that we have $\sigma(\sqrt{d_{n+1}}) = \epsilon\sqrt{d_{n+1}}$ with $\epsilon = \pm 1$.
(5) Show that we have $\sqrt{d_{n+1}} \in K_{n-1}$ if $\epsilon = 1$ and $\sqrt{d_{n+1}d_n} \in K_{n-1}$ if $\epsilon = -1$.
     Deduce a contradiction and conclude by induction.
(6) Show that the map

$$\begin{cases} \mathrm{Gal}(K_n/\mathbf{Q}) \to & \{\pm 1\}^n \\ \sigma \mapsto & (\sigma(\sqrt{d_1})/\sqrt{d_1}, \ldots, \sigma(\sqrt{d_n})/\sqrt{d_n}) \end{cases}$$

is an isomorphism of groups.
(7) Show that $\mathrm{Gal}(K_n/\mathbf{Q})$ has a (unique) structure of an $\mathbf{F}_2$-vector space compatible with its group structure. What is its dimension?
(8) How many subfields of degree 2 over $\mathbf{Q}$ does K_n have?

Exercise 12.3

(1) What is the Galois group of $\mathbf{Q}[\exp(\frac{21\pi}{35})]/\mathbf{Q}$? Is it cyclic?
(2) How many subfields of degree 12 does $\mathbf{Q}[\exp(\frac{21\pi}{35})]$ have? Of degree 6?

Exercise 12.4 Let k be a field and let $z_1, \ldots, z_n \in \bar{k}$ be the roots of a separable and monic polynomial $P \in k[X]$ of degree n with splitting field $K = D(P)$. We recall that the action of $\mathrm{Gal}(K/k)$ on the roots identifies $\mathrm{Gal}(K/k)$ with a subgroup of S_n. p, q denote two distinct odd prime numbers.

(1) Show that we have

$$\mathrm{disc}(P) = (-1)^{\frac{n(n-1)}{2}} \prod_{k=1}^{n} P'(z_k)$$

where $\mathrm{disc}(P)$ is the discriminant of P. Deduce the formula

$$q^* \overset{\mathrm{def}}{=} \mathrm{disc}(X^q - 1) = (-1)^{\frac{q(q-1)}{2}} q^q.$$

(2) We assume that the characteristic of k is either zero or an odd number different from q. Show that $X^q - 1$ is separable over k. Show that $\mathrm{Gal}(D(X^q - 1)/k)$ is contained in A_q if and only if q^* is a square in k.

We define the complex number $\zeta \overset{\mathrm{def}}{=} \exp(\frac{21\pi}{q})$ and let $G = \mathrm{Gal}(\mathbf{Q}[\zeta]/\mathbf{Q})$, which we identify with the cyclic group $(\mathbf{Z}/q\mathbf{Z})^*$ as in the course.

(3) Show that G is never contained in the alternating group A_q.

(4) Show that there exists a unique integer $\left(\frac{g}{q}\right)$ from $\{\pm 1\}$ such that $g^{\frac{q-1}{2}} = \left(\frac{g}{q}\right)$ mod q for all $g \in G = (\mathbf{Z}/q\mathbf{Z})^*$. By abuse, we will use the notation $\left(\frac{p}{q}\right)$ for $\left(\frac{p \bmod q}{q}\right)$.

Let $H \stackrel{\text{def}}{=} \{g \in G \text{ such that } g^{\frac{q-1}{2}} = 1\}$.

(5) Show that H is the unique subgroup of G of index 2 [first show that G is cyclic].
(6) Let γ be the morphism from G to G defined by $\gamma(g) = g^2$. What is the kernel of γ? Show that the image of γ is contained in H. Deduce the equality $H = \{g^2, g \in G\}$.
(7) What is the kernel of the signature $\epsilon : G = (\mathbf{Z}/q\mathbf{Z})^* \to S_q \to \{\pm 1\}$? Deduce the formula $\epsilon(g) = g^{\frac{q-1}{2}}$.
(8) Verify that there exists a unique $\Phi \in G$ such that $\Phi(\zeta) = \zeta^p$.
(9) Show that $\Phi \in H$ if and only if $p^{\frac{q-1}{2}} = 1$ mod q. Deduce the formula $\Phi(\sqrt{q^*}) = \left(\frac{p}{q}\right)\sqrt{q^*}$.

Let A be the ring $\mathbf{Z}[\zeta]$. We recall that there exists a prime ideal $\mathfrak{p}$ of A such that $\mathfrak{p} \cap \mathbf{Z} = p\mathbf{Z}$. Let $D_{\mathfrak{p}} \stackrel{\text{def}}{=} \{g \in G \text{ such that } g^{-1}(\mathfrak{p}) = \mathfrak{p}\}$ be the decomposition group of $\mathfrak{p}$.

(10) Recall why $\mathbf{F} \stackrel{\text{def}}{=} A/\mathfrak{p}$ is a finite field of characteristic p. Show by using the course that the canonical morphism $D_{\mathfrak{p}} \to \mathrm{Gal}(\mathbf{F}/\mathbf{F}_p)$ is bijective.
(11) Show that the image of Φ is the Frobenius morphism of $\mathrm{Gal}(\mathbf{F}/\mathbf{F}_p)$.
(12) Show that the action of $\mathrm{Gal}(\mathbf{F}/\mathbf{F}_p)$ on the $\bar{z}_i = z_i$ mod $\mathfrak{p}$ induces an embedding of $\mathrm{Gal}(\mathbf{F}/\mathbf{F}_p)$ into S_q, compatible (in a sense that will be specified) with the embedding of $D_{\mathfrak{p}}$ into S_q.
(13) Deduce that $\Phi \in A_q$ if and only if q^* is a square modulo p.
(14) Using the above, prove the quadratic reciprocity law

$$\left(\frac{p}{q}\right)\left(\frac{q}{p}\right) = (-1)^{\frac{(p-1)(q-1)}{4}}.$$

Exercise 12.5 Let $n \geq 2$ and m be natural numbers.

(1) Let $(a_1, \ldots, a_m)$ be an m-cycle of S_n and $\sigma \in S_n$. Show the formula

$$\sigma(a_1, \ldots, a_m)\sigma^{-1} = (\sigma(a_1), \ldots, \sigma(a_m)).$$

Let $s, \sigma \in S_n$ be cycles of respective lengths $n, n-1$ and (a, b), $1 \leq a < b \leq n$, a transposition. Let $\Sigma \subset S_n$ be the subgroup they generate.

(2) Show that there exists an $i \in \mathbf{N}$ such that $s^i \sigma s^{-i}$ fixes a.
(3) Show that $(a, a+1) \in \Sigma$ then that $(i, i+1) \in \Sigma$ for $1 \leq i < n$.

(4) Show that $\Sigma = S_n$.
(5) Let p be a prime number. Show that there exists an irreducible $P \in \mathbf{F}_p[X]$ of degree n.
(6) Show that there exists a monic polynomial $P \in \mathbf{Z}[X]$ and a prime number $p > 3$ such that

- The reduction of P modulo 2 is irreducible,
- The reduction of P modulo 3 is separable and has an irreducible factor of degree $n - 1$,
- The reduction modulo p is separable and has exactly $n - 2$ roots in $\mathbf{F}_p$.

(7) Show that any polynomial satisfying the conditions of the previous question admits S_n as its Galois group.

Exercise 12.6 Let $n \geq 3$ be an integer. Let $\zeta \in \mathbf{C}$ be a primitive n-th root of unity.

(1) Show the equality $[\mathbf{Q}[\zeta + \zeta^{-1}] : \mathbf{Q}] = \varphi(n)/2$.
(2) Show the equality $\mathbf{Q}[\zeta + \zeta^{-1}] = \mathbf{R} \cap \mathbf{Q}[\zeta]$.
(3) What is the image of complex conjugation in $(\mathbf{Z}/n\mathbf{Z})^*$ under the cyclotomic character?
(4) Show that $\mathbf{Q}[\zeta + \zeta^{-1}]/\mathbf{Q}$ is Galois and determine its Galois group.
(5) Let $x = \frac{a}{b} \in \mathbf{Q}$, where a and b are coprime. Calculate $[\mathbf{Q}[\cos(2\pi x)] : \mathbf{Q}]$ in terms of a and b.
(6) Suppose in this question $n \neq 4$. Show that we have

$$\left[\mathbf{Q}\left[\sin\left(\frac{2\pi}{n}\right)\right] : \mathbf{Q}\right] = \begin{cases} \varphi(n), & \text{if } \gcd(n, 8) < 4; \\ \varphi(n)/4 & \text{if } \gcd(n, 8) = 4; \\ \varphi(n)/2 & \text{if } \gcd(n, 8) > 4. \end{cases}$$

(7) Show that the area of a triangle in $\mathbf{R}^2$ with vertices in $\mathbf{Q}^2$ is a rational number.
(8) Show that squares with vertices in $\mathbf{Q}^2$ are the only regular polygons in $\mathbf{R}^2$ with vertices in $\mathbf{Q}^2$.
(9) Show that $I \in \mathbf{Q}[\zeta]$ if and only if n is a multiple of 4.
(10) Suppose n is not a multiple of 4. Show the equality $\mathbf{Q}[\sin(\frac{2\pi}{n}), I] = \mathbf{Q}[\zeta, I]$. Deduce $\mathbf{Q}[\zeta] = \mathbf{Q}[I \sin(\frac{2\pi}{n})]$.

We now suppose n is a multiple of 8.

(11) Show that we have $\sin(\frac{2\pi}{n}) \in \mathbf{Q}[\cos(\frac{2\pi}{n})]$. Deduce the equality $\mathbf{Q}[\sin(\frac{2\pi}{n})] = \mathbf{Q}[\cos(\frac{2\pi}{n})]$.
(12) Calculate the conjugates of $\sin(\frac{2\pi}{n})$ over $\mathbf{Q}$.

Exercise 12.7 Let $j : A \to B$ be a ring morphism. We further assume that B is an A-*module of finite type*. In other words, there exist $b_1, \ldots, b_n$ in B such that every element of B is a linear combination with coefficients in A of the b_i.

(1) Show that every element of B is integral over A.

We now assume A is non-zero, B is a field and j is injective, which allows us to consider A as a subring of B.

(2) Let $a \in A - \{0\}$. Show that there exists a $P \in A[X]$ such that $a^{-1} = P(a) \in B$. Deduce that $P(a) \in A$ is the inverse of a in A.
(3) Show that A is a field.

We propose to demonstrate the following result by induction on n. Let $B = k[x_1, \ldots, x_n]$ be a k-algebra integral of finite type over a field k. If B is a field, then $\dim_k(B) < \infty$.

(4) Prove the converse statement.

We assume $n \geq 1$ and that the theorem is proven for all field extensions $K[\xi_1, \ldots, \xi_{n-1}]/K$. We assume until (9) (inclusive) that x_n is transcendental over k and that $B = k[x_1, \ldots, x_n]$ is a field.

(5) Show that the k-algebra $k[x_n] \hookrightarrow B$ is isomorphic to a polynomial algebra over k and that the embedding $k[x_n] \to B$ extends uniquely to an embedding $k(x_n) = \mathrm{Frac}(k[x_n]) \hookrightarrow B$.
(6) Show that B is of finite dimension over $k(x_n)$.
(7) Show that there exists a $P \in k[x_n] - \{0\}$ such that x_i, $i = 1, \ldots, n$, is integral over the sub-ring $k[x_n, P^{-1}] \subset k(x_n)$.
(8) Show that B is a *module of finite type* over $k[x_n, P^{-1}]$. Deduce that $k[x_n, P^{-1}]$ is a field.
(9) By considering $x \in \bar{k}$, not a root of the polynomial P, show that there exists a unique surjective morphism of k-algebras $k[x_n, P^{-1}] \to k[x]$ that sends x_n to x. Show that the kernel is a non-zero ideal.
(10) Deduce that the hypothesis that x_n is transcendental over k is absurd.
(11) Prove the announced theorem.
(12) Show that the maximal ideals of the polynomial algebra $C[X_1, \ldots, X_n]$ are exactly the ideals $I_x = \{P \in C[X_1, \ldots, X_n]$ such that $P(x) = 0\}$, where $x = (x_1, \ldots, x_n) \in C^n$. Show that I_x is also the ideal generated by the monomials $X_i - x_i$, $i = 1, \ldots, n$.

Exercise 12.8 Let $k(t)$ be the field of rational fractions with coefficients in a field k. Let $n, m > 1$ be integers. We denote by ϖ the least common multiple of n, m and by δ their greatest common divisor.

(1) Show that the equation $x^n = t$ has no solution $x \in k(t)$ [one can write $x = P/Q$ with $\mathrm{GCD}(P, Q) = 1$].

Let K be the field $C(t)$ and Ω be an algebraic closure of K and denote by K_n the extension of K generated by *one* nth root $\sqrt[n]{t}$ of t in Ω.

(2) Show that the extension K_n/K is Galois and calculate its Galois group.
(3) How many subfields does K_n have that contain K? Can you identify them?

Let $H = \mathrm{Gal}(K_{nm}/K)$.

(4) Show that the extension K_{nm}/K_n is Galois and compare its Galois group H_n with H.

Let $K_n K_m$ be the subfield of K_{nm} generated by K_n and K_m.

(5) Show that the extension $K_{nm}/K_n K_m$ is Galois and calculate its Galois group in terms of H_n, H_m and H.
(6) Compare $K_n K_m$ and K_ϖ.
(7) Show that the extension $K_{nm}/K_n \cap K_m$ is Galois and calculate its Galois group in terms of H_n, H_m and H.
(8) Compare $K_n \cap K_m$ and K_δ.

Exercise 12.9 Let $n > 0$ be an integer and ℓ a prime number. We will say that an integer m is not an ℓ-power in a subring A of **C** if the equation $x^\ell = m$ has no solution $x \in$ A. We assume that n is not an ℓ-power in **Z**.

Let $\zeta = \exp(\frac{2i\pi}{\ell})$ and K be the splitting field over **Q** of $P(x) = X^\ell - n$.

(1) Show the equality $K = \mathbf{Q}(\zeta, \sqrt[\ell]{n})$.
(2) Let $x, y \in \mathbf{Q}$ such that $x^{\ell-1} = y^\ell$. Show that x is an ℓ-power in **Q**. If in addition x is an integer, show that x is an ℓ-power in **Z**.
(3) Show that n is not an ℓ-power in $\mathbf{Q}[\zeta]$ [Hint: let $N_{\mathbf{Q}[\zeta]/\mathbf{Q}}(n)$ be the determinant of the **Q**-linear map $n\mathrm{Id}_{\mathbf{Q}[\zeta]}$. Evaluate $N_{\mathbf{Q}[\zeta]/\mathbf{Q}}(n)$ in two different ways].
(4) Deduce that P is irreducible over $\mathbf{Q}[\zeta]$.

Let G be the Galois group of K over **Q**.

(5) Show that we have an exact sequence

$$0 \to \mathbf{Z}/\ell\mathbf{Z} \to G \to (\mathbf{Z}/\ell\mathbf{Z})^* \to 1.$$

Exercise 12.10 Let $M \in M_n(k)$ be a matrix with coefficients in an algebraically closed field k. We "recall" that there exists a *unique* pair of matrices $(D, N) \in (M_n(k))^2$ with D diagonalizable, N nilpotent, $DN = ND$ and $M = D + N$ (Jordan–Chevalley decomposition).

In the following two questions, $k = \mathbf{C}$.

(1) Let $M = \begin{pmatrix} 1 & 2 \\ 0 & z \end{pmatrix}$, $z \in \mathbf{C}$. Calculate D, N as above depending on the values of z.
(2) Deduce that the maps $M \mapsto D$ and $M \mapsto N$ are not continuous.

Let $M \in M_n(\mathbf{Q})$ and (D, N) be its Jordan–Chevalley decomposition in **C** where we regard M as a complex matrix.

(3) Show that M, $N \in M_n(\overline{\mathbf{Q}})$.
(4) Show that there exists a finite Galois extension $K/\mathbf{Q}$ such that D, $N \in M_n(K)$.

Let G be the Galois group of such an extension. We denote by $g(N), g(D)$ the matrices $[g(n_{i,j})]$, $[g(d_{i,j})]$ where $n_{i,j}$, $d_{i,j}$ are the coefficients of D, N.

(5) Show that D, N are fixed by G. Deduce that D, N $\in M_n(\mathbf{Q})$.
(6) Generalize the previous result to the case of matrices with coefficients in a perfect field.
(7) Let $k = \mathbf{F}_2(t)$ and $M = \begin{pmatrix} 0 & t \\ 1 & 0 \end{pmatrix}$. Show that the matrices D, N $\in M_2(\bar{k})$ of the Chevalley decomposition of M do not have coefficients in k.

Exercise 12.11 Let $P = X^3 + pX + q$, $p, q \in \mathbf{Q}$ and $K \subset \mathbf{C}$ be the extension of $\mathbf{Q}$ generated by its complex roots (possibly equal) z_1, z_2, z_3 of P. Let $G = \mathrm{Gal}(K/\mathbf{Q})$.

(1) Show the formula $\mathrm{disc}(P) = -\prod_i P'(z_i)$. Deduce the formula $\mathrm{disc}(P) = -4p^3 - 27q^2$.
(2) Show that P is reducible over $\mathbf{Q}$ if and only if it has a root in $\mathbf{Q}$. If P is reducible, determine G based on the number of rational roots.

We now assume P has no root in $\mathbf{Q}$.

(3) Show that the roots of P are simple.

We embed G in S_3 by making it act on its roots.

(4) What are the subgroups of S_3 ?
(5) Determine G based on the values of $\mathrm{disc}(P)$.
(6) Determine all the subfields of K according to the values of $\mathrm{disc}(P)$.
(7) Show that P is irreducible over $\mathbf{Q}(\sqrt{\mathrm{disc}(P)})$.

Exercise 12.12 Let n be an integer ≥ 1 and G the additive group $\mathbf{Z}/n\mathbf{Z}$.

(1) Prove that for every positive divisor d of n, there exists a unique subgroup G_d of cardinality d. Show that G_d is cyclic and exhibit a generator.
(2) Show that G_d is normal in G and show that the quotient is cyclic. Specify the cardinality of G/G_d.

Let $K/\mathbf{Q}$ be a Galois extension of the group G and $x \in K$ an element generating the extension. We denote by P the minimal polynomial of x (over $\mathbf{Q}$).

(3) Justify the existence of x. What remarkable property does the polynomial P have?
(4) Show that, depending on the parity of n, the field K has a unique subfield L such that $[K : L] = 2$ (resp. $[L : \mathbf{Q}] = 2$) or none.
(5) Show that K/L and L/$\mathbf{Q}$ are Galois and calculate the corresponding Galois groups.

Let $\sigma \in \mathrm{Aut}(\mathbf{C})$ denote complex conjugation.

(6) Show the equality $\sigma(K) = K$.

Assume in the following two questions that n is odd.

(7) Show that the discriminant of P is the square of a rational number.
(8) Show that the restriction of σ to K is the identity.

Assume n is even, $K \not\subset \mathbf{R}$ and let L' be the subfield of K fixed by σ.

(9) Show that we have $L = L'$ and $L \subset \mathbf{R}$ (cf. question 4).
(10) Deduce that if $m \in \mathbf{Q}$ satisfies $\sqrt{m} \in K$, then $m \geq 0$.

Exercise 12.13 Let n be an integer ≥ 1 and $\zeta = \exp(\frac{2i\pi}{n})$. Let p be a prime number not dividing n. For any field k, we denote by $\mu_n(k)$ the multiplicative group of n-th roots of 1 in k. We denote by $(\mathbf{Z}/n\mathbf{Z})^*$ the multiplicative group of invertible elements of the ring $\mathbf{Z}/n\mathbf{Z}$.

(1) Show that $X^n - 1 \in \mathbf{F}_p[X]$ has simple roots in $\overline{\mathbf{F}}_p$.

Let $K = \mathbf{Q}[\zeta]$, $A = \mathbf{Z}[\zeta]$ and $\mathfrak{p}$ a maximal ideal of A containing p. We denote by κ the finite field $\kappa(\mathfrak{p}) = A/\mathfrak{p}$.

(2) Show that the mapping $\xi \mapsto \bar{\xi} = (\xi \bmod \mathfrak{p})$ defines an isomorphism of groups $\mu_n(K) \to \mu_n(\kappa)$ and that these groups are cyclic of order n. Show that $\bar{\zeta}$ is of order n.
(3) Recall why $K/\mathbf{Q}$ and $\kappa/\mathbf{F}_p$ are Galois. Prove that the data of $\zeta, \bar{\zeta}$ define embeddings $G = \mathrm{Gal}(K/\mathbf{Q}) \subset (\mathbf{Z}/n\mathbf{Z})^*$ and $\overline{G} = \mathrm{Gal}(\kappa/\mathbf{F}_p) \subset (\mathbf{Z}/n\mathbf{Z})^*$.

We now identify $G, \overline{G}$ with subgroups of $(\mathbf{Z}/n\mathbf{Z})^*$.
Let $D(\mathfrak{p})$ be the subgroup of G fixing $\mathfrak{p}$.

(4) Show that the morphism $D(\mathfrak{p}) \to G \to (\mathbf{Z}/n\mathbf{Z})^*$ identifies with the composition of the natural arrows $D(\mathfrak{p}) \to \overline{G} \to (\mathbf{Z}/n\mathbf{Z})^*$.
(5) Show that there exists a unique $F_{\mathfrak{p}} \in D(\mathfrak{p})$ whose image is the Frobenius morphism in $\overline{G}$. Show that the image of $F_{\mathfrak{p}}$ in $G \subset (\mathbf{Z}/n\mathbf{Z})^*$ is $p \bmod n$.
(6) Deduce $G = (\mathbf{Z}/n\mathbf{Z})^*$.
(7) Deduce from the previous question that the cyclotomic polynomial $\Phi_n(X) = \prod_{\substack{(m,n)=1 \\ 1 \leq m \leq n}} (X - \exp(\frac{2i\pi}{n}))$ is irreducible over $\mathbf{Q}$ (without using the proof from the course).

Exercise 12.14

(1) Show, for example by a direct computation, that $5 + \sqrt{21}$ is not a square in $\mathbf{Q}[\sqrt{21}]$ [set $5 + \sqrt{21} = (a + b\sqrt{21})^2$, $a, b \in \mathbf{Q}$ and search for possible values for a^2].

Let $z = \sqrt{5 + \sqrt{21}}$ and $K = \mathbf{Q}[z]$.

(2) Show $[K : \mathbf{Q}] = 4$.
(3) Let $z' = \sqrt{5 - \sqrt{21}}$. Show $z' \in K$ [calculate zz']. Deduce the conjugates (over $\mathbf{Q}$) of z and then that $K/\mathbf{Q}$ is Galois.

Let $G = \mathrm{Gal}(K/\mathbf{Q})$.

(4) Show that there exists a unique element $g \in G$ such that $g(z) = -z$.
(5) Show that there exists a unique element $h \in G$ such that $h(z) = z'$.

(6) Show that we have $g(z') = -z'$ and $h(z') = z$. Deduce that g and h commute and that $G \simeq (\mathbf{Z}/2\mathbf{Z})^2$.

(7) Describe the subfields of K (specify a primitive element for each of the corresponding extensions of **Q**).

(8) Show (preferably without calculation) that z cannot be written without using radicals.

(9) Can we write z as a fourth root of a rational number?

Exercise 12.15 Let $P(X) = \prod_{i=1}^{n}(X - x_i)$ be a polynomial of $\mathbf{Q}[X]$ of degree $n \geq 1$ and let $y_1, \ldots, y_{n-1} \in \mathbf{C}$ be the complex roots (possibly equal) of $P'(X)$.

(1) Show the formula

$$\mathrm{disc}(P) = (-1)^{\frac{n(n-1)}{2}} \prod_{i=1}^{n} P'(x_i).$$

From this, deduce the formula

$$\mathrm{disc}(P) = n^n (-1)^{\frac{n(n-1)}{2}} \prod_{i=1}^{n-1} P(y_i).$$

(2) Show that the discriminant of $X^n + aX + b$ is

$$(-1)^{\frac{n(n-1)}{2}} (1-n)^{n-1} a^n + n^n b^{n-1}).$$

Let $P(X) = X^5 + 20X + 16$ and G be its Galois group over **Q**.

(3) Show that the complex roots of P are simple.

We choose a numbering of the complex roots of P defining an embedding of G in S_5.

(4) Show that P has a unique real root. Deduce that G contains a double transposition.

(5) Show that G is contained in A_5.

(6) Factorize $P(X)$ mod $7 \in \mathbf{F}_7[X]$ into irreducible factors [observe that -2 and -3 are the only roots in $\mathbf{F}_7$]. Deduce that G contains a 3-cycle.

Let $\overline{P} = P \bmod 3 \in \mathbf{F}_3[X]$.

(7) Show that $\overline{P}$ has no root in $\mathbf{F}_9$ [observe that for all $x \in \mathbf{F}_9^*$ we have $x^5 = \pm x$].

(8) Deduce that $\overline{P}$ is irreducible then that G contains a 5-cycle.

(9) Show that a 3-cycle and a double transposition of S_4 generate A_4.

(10) Show that a 5-cycle, a 3-cycle and a double transposition of S_5 generate A_5.

(11) Show that $G = A_5$.

Exercise 12.16

(1) Let G be a finite cyclic group and δ a divisor of card(G). Show that $G_\delta = \{g^\delta, g \in G\}$ is the unique subgroup of G of cardinal card(G)$/\delta$.

Let p be a prime number > 2 and consider the cyclotomic extension of K/Q (contained in C/Q) with

$$K = Q[\zeta], \quad \zeta = \exp\left(\frac{2I\pi}{p}\right).$$

Let $G = \text{Gal}(K/Q)$.

(2) Show that K/Q is Galois with a cyclic Galois group isomorphic to $(Z/pZ)^*$.

Let d be a divisor of $p - 1$.

(3) Show that K contains a unique subfield K_d of degree d over Q.
(4) Show that K_d is Galois over Q and that we have $K^{G_d} = K_d$.
(5) Every element of G can be regarded as an element of the space $\text{End}_Q(K)$ of Q-linear endomorphisms of K. Let

$$p_d = \frac{d}{p-1} \sum_{g \in G_d} g \in \text{End}_Q(K).$$

Show the formula

$$g p_d = p_d g = p_d$$

for all $g \in G_d$ and then that p_d is a projector of image K_d.

Let

$$\zeta_d = \sum_{k=0}^{p-1} \zeta^{k^d}.$$

(6) Compare ζ_d and $p_d(\zeta)$. Deduce that $Q[\zeta_d] \subset K_d$.
(7) Show that $\{g(\zeta), g \in G\}$ is a Q-basis of K.
(8) Show $p_d(g(\zeta)) \in Q[\zeta_d]$ for all $g \in G$ [observe that we have $g p_d = p_d g$ for all $g \in G$]. Deduce that $K_d = Q[\zeta_d]$.
(9) Show $K_{\frac{p-1}{2}} = Q[\cos(\frac{2\pi}{p})]$.
(10) Show $K_2 = Q[\sqrt{\epsilon p}]$ with $\epsilon = 1$ if -1 is a square modulo p and $\epsilon = -1$ otherwise.

Exercise 12.17 Let p be an odd prime number. We identify the symmetric group S_p with the bijections of Z/pZ and we denote by $\bar{m}$ the class of the integer m in

$\mathbf{Z}/p\mathbf{Z}$. Let i, j be two integers with $1 \le i < j \le p$ and G the subgroup of S_p generated by the cycle $(\bar{1}, \bar{2}, \ldots, \bar{p})$ and the transposition $(\bar{i}, \bar{j})$.

(1) Show that for all $k \in \mathbf{Z}$, we have $(\bar{i} + \bar{k}, \bar{j} + \bar{k}) \in G$ and then that

$$(\bar{i} + k\overline{(j - i)}, \bar{i} + (k + 1)\overline{(j - i)}) \in G.$$

(2) Show by induction on $k \in [1, \ldots, p - 1]$ that

$$(\bar{i}, \bar{i} + k\overline{(j - i)}) \in G.$$

(3) Show that the equation $\bar{i} + \bar{k}\overline{(j - i)} = \bar{i} + 1$ has a solution $\bar{k} \in (\mathbf{Z}/p\mathbf{Z})^*$.
(4) Show that $(\bar{i}, \bar{i} + 1) \in G$ and then

$$\forall \bar{t} \in \mathbf{Z}/p\mathbf{Z}, \ (\bar{t}, \bar{t} + 1) \in G.$$

(5) Show $G = S_p$.
(6) Let c be a p-cycle and τ a transposition of S_p. Show that S_p is generated by c and τ.
(7) Show that the previous result fails if we do not assume p is prime.

Exercise 12.18 Let k be a field and $P \in k[X]$ of degree $n > 0$.

(1) Show that if P is reducible, it has a root in an extension of k of degree $\le n/2$.

Let p be a prime number and $P \in \mathbf{F}_p[X]$ of degree $n > 0$.

(2) Show that P is irreducible if and only if it has no root in the fields $\mathbf{F}_{p^d}, d \le n/2$.
(3) Conclude from the previous question that P is irreducible if and only if

$$\forall d, \ 1 \le d \le n/2, \ \mathrm{GCD}(P, X^{p^d} - X) = 1.$$

Exercise 12.19 Let $P = X^5 - X + 3 \in \mathbf{Q}[X]$ and G its Galois group over $\mathbf{Q}$.

(1) Show that P is separable. Deduce that G embeds into S_5.
(2) Factorize P in $\mathbf{F}_3[X]$
(3) Show that P is irreducible in $\mathbf{F}_5[X]$. [You may use a previous exercise.]
(4) Show that G is isomorphic to S_5. [You may use a previous exercise.]
(5) Show that P is irreducible over $\mathbf{Q}$.
(6) Does there exist a positive integer $n > 0$ such that at least one root of P is contained in $\mathbf{Q}[\exp(\frac{2i\pi}{n})]$?

Exercise 12.20 We equip $\mathbf{C}$ with the structure of an oriented Euclidean plane in the trigonometric sense. Let C be the set of the 4 vertices of a square and ω its center. We denote by Γ the subgroup of bijections g of C such that

$$\forall x, y \in \mathbf{C}, \ |g(x) - g(y)| = |x - y|.$$

Let ρ be the rotation of center ω and angle $\frac{\pi}{2}$ and σ a symmetry with respect to a diagonal of C (or rather their restrictions to C).

(1) Show that if $g \in \Gamma$ fixes two consecutive vertices of C, then $g = \text{Id}$.
(2) Show the equality

$$\Gamma = \{\rho^\alpha, \rho^\beta\sigma, \ \alpha, \beta \in [1, \dots, 4]\}$$

and that we have the formula $\sigma\rho\sigma = \rho^{-1}$.
(3) Show that Γ is a non-abelian group of order 8 and that we have an exact sequence of groups

$$1 \rightarrow \mathbf{Z}/4\mathbf{Z} \rightarrow \Gamma \rightarrow \mathbf{Z}/2\mathbf{Z} \rightarrow 1.$$

(4) Give all the subgroups of Γ. Which ones are normal? In particular, how many subgroups of order 2 does Γ have? 4?

Let G be the Galois group over $\mathbf{Q}$ of $X^4 - 2$, that is to say of K/$\mathbf{Q}$ where K is the subfield of $\mathbf{C}$ generated by the set C of complex roots of $X^4 - 2$. We set $x = 2^{1/4}$.

(5) Show that L $= \mathbf{Q}[x]$ is non-Galois over $\mathbf{Q}$. Deduce that G is a non-commutative group.
(6) Show K $= \mathbf{Q}[x, I]$ and $[K : \mathbf{Q}] = 8$. Deduce the cardinality of G.
(7) Show that the action of G on C induces an isomorphism of G onto Γ.
(8) Show that there exists a unique element $r \in$ G such that $r(x) = Ix$ and $r(I) = I$. What is the order of r?
(9) Show that there exists a unique element $s \in$ G such that $s(I) = -I$ and $s(x) = x$. What is the order of s?
(10) Show the formula $srs = r^{-1}$. Also show that s and r generate G.
(11) Find all the subfields of K. Which ones are Galois over $\mathbf{Q}$? In particular, how many subfields of degree 2 are there? 4?
(12) Give a primitive element of K over $\mathbf{Q}$.
(13) Let K/k be an algebraic extension of perfect fields. We assume that every non-constant polynomial with coefficients in k has at least one root in K. Show that K is an algebraic closure of k.

Exercise 12.21 We consider the polynomial

$$P(X) = X^5 - 5X^2 + 1 \in \mathbf{Q}[X]$$

and G the Galois group of P over $\mathbf{Q}$.

(1) Show that P is separable and that there exists an injective group morphism $\phi :$ G $\rightarrow S_5$.

Let $\overline{P}$ be the reduction of P in $\mathbf{F}_2[X]$.

(2) Show that $\overline{P}$ has no root in $\mathbf{F}_2$.

(3) Show that $\overline{P}$ has no root in $\mathbf{F}_4$.
(4) Deduce that $\overline{P}$ is irreducible in $\mathbf{F}_2[X]$.
(5) Deduce that P is irreducible in $\mathbf{Z}[X]$, and in $\mathbf{Q}[X]$.
(6) Show that 5 divides the order of G without using the theorem of reduction modulo p.
(7) Show that an element of S_5 of order 5 is a 5-cycle.
(8) Using a reduction modulo p, show that G has an element of order 5.
(9) How many real roots does P have?
(10) Let σ be the complex conjugation. Show that $\sigma \in G$ and determine the type of $\phi(\sigma)$ in S_5.

We admit that an arbitrary transposition and 5-cycle generate S_5.

(11) Calculate $|G|$. Is the equation $P(x) = 0$ solvable by radicals?

Exercise 12.22 Let k be a perfect field of characteristic $p \neq 2$ and $\overline{k}$ an algebraic closure of k.

Let $m \geq 1$ be an integer coprime with p if $p \neq 0$. We assume that k contains a primitive m^{th} root of 1, ϵ.

Let $a \in k$ be fixed. We consider the polynomial

$$P(X) = X^{2m} - 2aX^m + 1.$$

Let K be the splitting field of P in $\overline{k}$.

(1) Recall why the extension K/k is Galois.
(2) Determine K for $a = 1$.
(3) Show that at least one of the roots of $X^2 - \epsilon$ in $\overline{k}$ is an mth root of -1.
(4) Determine K for $a = -1$. You may distinguish between the two cases according to whether or not k contains an mth root of -1, and use the polynomial $X^2 - \epsilon$.
(5) Find a condition on the element a which is equivalent to P being separable.
(6) Give the solutions of the equation $P(x) = 0$ in $\overline{k}$ (you may use a, ϵ and radicals).
(7) Show that $G = \text{Gal}(K/k)$ is a solvable group.

We fix $x \in K$ such that $P(x) = 0$.

(8) Show that $K = k[x]$ and that $[K : k] \leq 2m$.

We assume in the rest of this exercise that $[K : k] = 2m$.

(9) Show that P is irreducible in $k[X]$.

(10) Show that the conjugates of x over k in $\overline{k}$ are the $x\epsilon^j$, $x^{-1}\epsilon^j$ with $0 \leq j < m$.
(11) Show that there exists a unique $g \in G$ such that $g(x) = \epsilon x$ and describe a diagonalization basis of g.
(12) Show that G has a cyclic subgroup H of order m.
(13) Describe K^H (you may give a primitive element of the extension K^H/k). Show that H is normal in G.

(14) Show that G is involved in an exact sequence

$$1 \to \mathbf{Z}/m\mathbf{Z} \to G \to \mathbf{Z}/2\mathbf{Z} \to 1.$$

(15) Show that there exists a unique $h \in G$ such that $h(x) = x^{-1}$. Show that the subgroup H' of G generated by h is of order 2. Describe $K^{H'}$ (you may give a basis and a primitive element).
(16) Show that H' is normal in G if and only if $m \le 2$.
(17) Show that G is commutative if and only if $m \le 2$.

Exercise 12.23 We propose to demonstrate the following result:
  For all $a \in \mathbf{Q}$, there exists an $n \in \mathbf{N}$ such that $\mathbf{Q}(\sqrt{a}) \subset \mathbf{Q}(e^{2\mathrm{I}\pi/n})$.

(1) Directly handle the cases $a = -1$ and $a = 2$.

We now fix an odd prime number p, and we set $K = \mathbf{Q}(e^{2\mathrm{I}\pi/p})$. We recall that K is Galois over $\mathbf{Q}$ and that its Galois group G is identified with $(\mathbf{Z}/p\mathbf{Z})^* \simeq \mathbf{Z}/(p-1)\mathbf{Z}$. We also set $\xi = e^{2\mathrm{I}\pi/p}$.

(2) Show that

$$\xi + \xi^2 + \cdots + \xi^{p-1} = -1$$

and that $\mathcal{B} = \{\xi, \xi^2, \ldots, \xi^{p-1}\}$ is a basis of the $\mathbf{Q}$-vector space K.
(3) Deduce that if $z = \sum_{1 \le i < p} a_i \xi^i \in \mathbf{Q}$ with $a_i \in \mathbf{Q}$ then

$$a_1 = a_2 = \cdots = a_{p-1} \text{ and } z = -a_1.$$

(4) Denote by $H = \{1^2, 2^2, \ldots, (p-1)^2\} \subset (\mathbf{Z}/p\mathbf{Z})^*$ the set of squares of G. Show that H is a subgroup of G of order $(p-1)/2$.
(5) Using $\mathcal{B}$, show that $x := \sum_{g \in H} g(\xi) \notin \mathbf{Q}$. Using the Galois correspondence, deduce that $[\mathbf{Q}[x] : \mathbf{Q}] = 2$.
(6) Denote by $\overline{H} \subset G$ the complement of H in G. Show that if $g \in \overline{H}$ then $g \cdot H = \overline{H}$. Deduce that the conjugates of x over $\mathbf{Q}$ are x and $x' = \sum_{g \in \overline{H}} g(\xi)$.
(7) Show that $x + x' = -1$ then that $x^2 + x \in \mathbf{Q}$.

We now propose to explicitly calculate $x^2 + x$.

(8) Suppose that $-1 \in H$. By expanding in $\mathcal{B}$, show that

$$x^2 + x = \frac{p-1}{2} + \sum_{1 \le i < p} a_i \xi^i$$

where the $a_i \in \mathbf{Q}$ satisfy $\sum_{1 \le i < p} a_i = (p-1)^2/4$. Deduce, with the help of 3), the value of $x^2 + x$ and that $\mathbf{Q}(x) = \mathbf{Q}(\sqrt{p})$.

(9) We now suppose that $-1 \in \overline{H}$. Show similarly that

$$x^2 + x = \sum_{1 \le i < p} a_i \xi^i$$

where the $a_i \in \mathbf{Q}$ satisfy $\sum_{1 \le i < p} a_i = (p-1)(p+1)/4$. Deduce the value of $x^2 + x$ and that $\mathbf{Q}(x) = \mathbf{Q}(\sqrt{-p})$.

(10) Conclude.

Chapter 13
Solutions to Exercises

In this chapter, we sketch the solutions to the most typical exercises of Chaps. 2–11 and then give detailed solutions to the review exercises of Chap. 12.

Brief Solution to Exercise 2.4.5
The surjective morphism det : $\mathrm{GL}_n(\mathbf{C}) \to \mathbf{C}^*$ sends $\mathrm{SL}_n(\mathbf{C})$ to $\{1\}$ and therefore induces a surjection $\delta : \mathrm{GL}_n(\mathbf{C})/\mathrm{SL}_n(\mathbf{C}) \to \mathbf{C}^*$. To say $\delta(\mathrm{M} \bmod \mathrm{SL}_n(\mathbf{C})) = 1$ is to say $\delta(\mathrm{M}) = 1$, proving that δ is injective, therefore δ is an isomorphism since we know it is surjective.

Brief Solution to Exercise 2.6.5
If $x = \sigma(a_i)$, we have

$$\sigma(a_1, \ldots, a_k)\sigma^{-1}(x) = \sigma(a_1, \ldots, a_k)(a_i) = \sigma(a_{i+1})$$

for $i \in \mathbf{Z}/k\mathbf{Z}$.

Let S be the group generated by the $(i, i+1)$. We deduce the formula

$$(i+1, j)(i, i+1)(i+1, j) = (i, j)$$

for all $j \neq i, i+1$ which shows (by induction) $(i, j) \in S$ for all j and therefore $S = S_n$ since the transpositions generate S_n.

Similarly, the formula $(1, \ldots, n)^j(1, 2)(1, \ldots, n)^j = (j, j+1)$ proves that $(1, 2)$ and $(1, \ldots, n)$ generate S_n according to what precedes.

Finally, suppose that a subgroup S contains an n-cycle, a transposition and an $(n-1)$-cycle. By renumbering if necessary, we can assume $c = (1, \ldots, n) \in S$. Let $t = (i, j), i < j$ be a transposition in S. By conjugating by c, we can assume $t = (1, j)$. Let γ be an $(n-1)$-cycle of S and let a be the unique point fixed by γ. By conjugating by c^{n-a+1}, we can assume $a = 1$. Then there exists a $d \in \mathbf{Z}$ such that $\gamma^d(j) = 2$ since γ induces an $(n-1)$-cycle of $S_{n-1} = \mathrm{Bij}(\{2, \ldots, n\})$.

But $\gamma^d(1, j)\gamma^{-d} = (1, 2)$ so that $(1, 2) \in S$ and the result follows by the previous point.

Brief Solution to Exercise 2.6.8

Suppose H has index 2. Let $g \in G$. We must show $gH = Hg$. If $g \in H$, it is clear. Otherwise, $gH \neq H$ and $Hg \neq H$. But G/H has cardinality 2, so it equals $\{H, gH\}$. Since G is the disjoint union of its right classes, we have $gH = G - H$. Similarly, $Hg = G - H$, and therefore $gH = Hg$. Let γ be the unique non-neutral element of the group G/H. There exists a unique isomorphism $G/H \overset{\sim}{\to} \{\pm 1\}$: it sends γ to -1. If $G = S_n$, as the transpositions are conjugate, their images in the *abelian* group $\{\pm 1\} = G/H$ are either always 1 or always -1. As the transpositions generate $G = S_n$, it cannot be 1 otherwise the quotient morphism would not be surjective. The image is therefore -1 and therefore the quotient morphism is the signature. Its kernel H is therefore A_n, which is what we wanted.

Brief Solution to Exercise 2.7.2

It is well known (use the Gauss pivot) that $SL_n(\mathbf{C})$ is generated by the transvections $T_{i,j}(\lambda) = I + \lambda E_{i,j}, i \neq j, \lambda \neq 0$ and that two transvections are conjugate in $SL_n(\mathbf{C})$. Therefore, there exists a $P \in SL_n(\mathbf{C})$ such that

$$(T_{i,j}(\lambda))^2 = T_{i,j}(2\lambda) = PT_{i,j}(\lambda)P^{-1}$$

so that $T_{i,j}(2\lambda)$ is the commutator $[P, T_{i,j}(\lambda)]$. As the transvections generate $SL_n(\mathbf{C})$, we have $D(SL_n(\mathbf{C})) = SL_n(\mathbf{C})$. Note that the argument is still valid if we replace $\mathbf{C}$ with any field of characteristic different from 2.

Brief Solution to Exercise 2.7.8

(1) It is enough to associate with an invertible triangular matrix the sequence of its diagonal coefficients to find the exact sequence sought. We then invoke Proposition 2.7.6.
(2) The first point is clear. Let $f \in U_j$ and $g \in U$ (or $g \in B$ if you want). Suppose that $Id + u$, $Id + v \in U_j$. We have

$$\ln((Id + u)(Id + v))(F_i) = (u + v + uv)(F_i) \subset F_{i-j} + uv(F_i).$$

Since

$$uv(F_i) \subset F_{i-2j} \subset F_{i-j},$$

we have indeed $(Id + u)(Id + v) \in U_j$. Since Id is in U_j, the latter is indeed a subgroup of U.

Since $g(F_i) \subset F_i$ and $g^{-1}(F_i) \subset F_i$ for all i, we have

$$gfg^{-1}(F_i) \subset gf(F_i) \subset g(F_{i-j}) \subset F_{i-j}.$$

Thus U_i is a normal subgroup of U.

(3) Let $j \geq 1$. Since $\ln(f)$ leaves F_i and F_{i-j-1} stable, it indeed induces a linear map $\ln(f)_{i,j}$ of the quotient F_i/F_{i-j-1}. To say that $\ln(f)_{i,j}$ is zero is equivalent to saying $\ln(f)(F_i) \subset F_{i-j-1}$.

(4) If, as above, $\mathrm{Id} + u$, $\mathrm{Id} + v \in U_j$, we have

$$uv(F_i) \subset F_{i-2j} \subset F_{i-j-1}$$

and therefore is zero as an endomorphism of F_i/F_{i-j-1}. This ensures that $f \mapsto \ln(f)_{i,j}$ is a morphism from U_i into the additive commutative group $\mathrm{End}(F_i/F_{i-j-1})$. According to (3), the kernel of $\ln_j$ is U_{j-1}.

(5) We apply Definition 2.7.1 to conclude that U is solvable and therefore that B is solvable according to (1).

Brief Solution to Exercise 2.7.11

The fact that S_4 operates on X results from Proposition 2.6.4 or formula (2.6.1) as you wish. If we number the elements of X by deciding that the transposition $(1, i + 1)$ appears in x_i, the operation $S_4 \to \mathrm{Aut}(X)$ is identified with a morphism $S_4 \to S_3$. The image of $(1, 2)$ is $(2, 3)$. We deduce that π is surjective. Since the group $K = \{\mathrm{Id}\} \cup X = \mathbf{Z}/2\mathbf{Z} \times \mathbf{Z}/2\mathbf{Z}$ generated by X is abelian, it is in the kernel. For reasons of cardinality, it is the kernel. Therefore $\mathrm{Ker}(\pi)$ is solvable because it is abelian and $S_4/\mathrm{Ker}(\pi) = S_3$ is solvable, so that S_4 is solvable (Proposition 2.7.6).

Brief Solution to Exercise 3.6.11

Let $a \in A$ have an invertible image in A/I. By definition, there therefore exist $b \in A$ and $i \in I$ such that $ab = 1 + i$. But the inverse of $1 + i$ is $\sum_{k=0}^{n-1}(-i)^k$ thanks to the formula for a geometric series. Thus, a is invertible with inverse $b/(1 + i)$.

Brief Solution to Exercise 3.8.3

Decompose n, d into prime factors:

$$n = \prod p^{n_p}, \; d = \prod p^{m_p}, m_p \leq n_p.$$

According to the Chinese Lemma, the morphism

$$(\mathbf{Z}/n\mathbf{Z})^* \to (\mathbf{Z}/d\mathbf{Z})^*$$

is surjective if and only if each morphism

$$(\mathbf{Z}/p^{n_p}\mathbf{Z})^* \to (\mathbf{Z}/p^{m_p}\mathbf{Z})^*$$

is. We can therefore assume $n = p^{n_p}$. But then, $\mathbf{Z}/p^{m_p}\mathbf{Z}$ is identified with the quotient of $\mathbf{Z}/p^{n_p}\mathbf{Z}$ by the ideal I generated by p^{m_p}. As $I^{n_p-m_p} = (0)$, we invoke Exercise 3.6.11.

Brief Solution to Exercise 4.1.8
According to the universal property of the quotient (Theorem 3.4.4), the **R**-algebra morphisms from **R**$[X]/(P(X))$ into an algebra A are identified with the roots of P in A. Let $j = \exp(\frac{2i\pi}{3})$. The morphism **R**$[X] \to$ **C** defined by $X \mapsto j$ passes to the quotient to give an **R**-linear morphism $K = $ **R**$[X]/(X^2 + X + 1) \to$ **C**, which is clearly surjective. It is injective for reasons of dimension, for example (or simply because it is a morphism of fields). Similarly, we have two morphisms of **R**-algebras **R**$[X]/(X(X+1)) \to$ **R** that send X to 0 and -1 respectively. The corresponding morphism **R**$[X]/(X(X+1)) \to$ **R**^{2} is clearly surjective, and injective for reasons of dimension.

Brief Solution to Exercise 4.7.4
If $P(l) = 0, l \in L$, there exists a unique k-algebra morphism from $k[X]/(P)$ into L that sends X to l (Theorem 3.4.4), hence the first point. If P is arbitrary and non-constant, let us show by induction on n that for any polynomial of degree $\leq n$ there exists an extension K of k of degree $\leq n!$ in which P is split. If $n = 0$, it is clear. Suppose $n > 0$ and the statement is true for $n - 1$. Write $P = P_1 P_2$ with P_1 irreducible. Let l be a root of P_1 in the splitting field L of P_1, which is of degree $\deg(P_1) \leq n$. We write $P_1 = (X-l)P_3$ with $P_3 \in L[X]$. We have $\deg(P_2 P_3) \leq n-1$. By induction, there exists an extension K of L of degree $\leq (n-1)!$ such that $P_2 P_3$ is split over K. We have $[K : k] \leq n!$ (4.2.4) and K is suitable.

Brief Solution to Exercise 5.2.5
Let $n > 0$ and μ_x be the minimal polynomial, necessarily irreducible, of a generator x of the multiplicative group of $\mathbf{F}_{q^n}^*$. As $\mathbf{F}_{q^n} = \mathbf{F}_q[x]$, we have $\deg(\mu_x) = n$, which is what we wanted. If P is irreducible of degree n, the rupture field is of degree n over $\mathbf{F}_q$. If x is a root of P in $\overline{\mathbf{F}}_p$, it is identified (Exercise 4.7.4) with $\mathbf{F}_q[x]$, which is $\mathbf{F}_{q^n}$ for reasons of dimension. It is independent of the root x so that all the roots of P are in $\mathbf{F}_{q^n}$. Thus, P is split over its rupture field $\mathbf{F}_{q^n}$, which is also its splitting field. As the roots of P are simple and in $\mathbf{F}_{q^n}$, the set of roots of $X^{p^n} - X$, we have $P | (X^{q^n} - X)$.

Brief Solution to Exercise 5.6.4
The elements $X^{1/p}, Y^{1/p}$ are algebraic of degree p ($T^p - X$ is irreducible in $\mathbf{F}_p(X, Y)[T]$ according to Lemma 5.4.5). The extension $k[X^{1/p}, Y^{1/p}]/k$ is in particular finite. The formula

$$P(X^{1/p}, Y^{1/p})^p = P_p(X, Y),$$

where $P_p(U, V) = \sum_{i,j} a_{i,j}^p U^i V^j$ with

$$P(U, V) = \sum_{i,j} a_{i,j} U^i V^j \in k[U, V],$$

ensures that every element of $k(X^{1/p}, Y^{1/p})$ is of degree at most p. If the extension in question was simple, it would be of degree p so that we would have $k[X^{1/p}] = k[Y^{1/p}]$ for reasons of dimension. We would therefore be able to write $X = \sum a_i(X^p, Y^p)Y^i$, where the a_i are rational fractions. Differentiating with respect to X, we get $1 = 0$, a contradiction.

Brief Solution to Exercise 8.3.6

Write $x = p/q$ with p, q coprime and $q \geq 1$. Then, x is a root of a polynomial of the form $P(X) = X^n + \sum_{i<n} a_i X^i$, $n \geq 1$ with $a_i \in \mathbf{Z}$. We therefore have

$$q^n P(p/q) = p^n + q \sum_{i<n} a_i p^i q^{n-1-i} = 0,$$

so that $q | p^n$. Since $GCD(p, q) = 1$, this forces $q = 1$ and therefore $x = p \in \mathbf{Z}$.

Brief Solution to Exercise 9.2.2

First of all, the Galois group G of a separable polynomial of degree d is contained in S_d and is trivial only if P is split over k (since the degree of the splitting field of P is the cardinality of the Galois group). This settles the degree 2 case. In degree 3, we can assume P has no root in k (otherwise we have a trivial group or $\mathbf{Z}/2\mathbf{Z}$ according to what precedes), hence it is irreducible here. The cardinality of the Galois group is therefore divisible by 3, and thus is either 3 or 6. However, S_3 has a unique subgroup of cardinality $3!/2 = 3$ (Exercise 2.6.8), the alternating group $A_3 = \mathbf{Z}/3\mathbf{Z}$. If k is of odd characteristic, this occurs precisely if disc(P) is a square in k (Proposition 9.2.1).

Brief Solution to Exercise 9.2.3

The derivative of $X^n - 1$ is nX^{n-1}, which has a non-zero root only if $p|n$. We immediately deduce that P is separable if and only if p and n are coprime.

In general, if $P \in k[X]$ is monic of degree n with roots $x_1, \ldots, x_n$ in $\bar{k}$, we have

$$disc(P) = (-1)^{\frac{n(n-1)}{2}} \prod_i \prod_{j \neq i} (x_i - x_j) = (-1)^{\frac{n(n-1)}{2}} \prod_i P'(x_i).$$

If $P = X^n - 1$, we therefore have

$$disc(P) = (-1)^{\frac{n(n-1)}{2}} n^n \prod_i x_i^{n-1} = (-1)^{\frac{n(n-1)}{2}} n^n \left(\prod_i x_i \right)^{-1}.$$

The product of the roots of $X^n - 1$ is $(-1)^{n-1}$, so that

$$disc(X^n - 1) = (-1)^{\frac{n(n+1)}{2}} n^n.$$

Suppose $GCD(p, n) = 1$ and $p \neq 2$. The action of $Gal(K/k)$ on the set $\mu_n(\bar{k})$ "is in A_n" if and only if $(-1)^{\frac{n(n+1)}{2}} n^n$ is not a square in k (Proposition 9.2.1).

Brief Solution to Exercise 9.2.4

Since k is of characteristic 2, we have

$$x_i^2 + x_j^2 = (x_i - x_j)^2 \neq 0$$

for all $i \neq j$.

Let $\mathcal{P}$ be the set of *pairs* $\pi = \{x, y\}$ where x, y are two distinct roots of P. The Galois group G of P permutes the pairs by acting on the roots. We denote by π_1, π_2 the elements xy and $x^2 + y^2$ respectively. We have $a = \sum_{\pi \in \mathcal{P}} \frac{\pi_1}{\pi_2}$, which is clearly invariant under g and is therefore an element of k.

We have

$$b^2 + b = \sum_{i<j} \left(\frac{x_i^2}{x_i^2 + x_j^2} + \frac{x_i(x_i + x_j)}{x_i^2 + x_j^2} \right) = \sum_{i<j} \frac{x_i x_j}{x_i^2 + x_j^2} = a.$$

The sum of the roots of $X^2 + X + a$ is 1 so its roots are b or $b + 1$. Since $X^2 + X + a \in k[X]$, the group G permutes its roots so that $g(b) = b$ or $b + 1$. If g acts on the x_i by the permutation σ of the indices, we have

$$g\left(\frac{x_i}{x_i + x_j} \right) = \frac{x_{\sigma(i)}}{x_{\sigma(i)} + x_{\sigma(j)}} = 1 + \frac{x_{\sigma(j)}}{x_{\sigma(i)} + x_{\sigma(j)}}.$$

From this we deduce the formula

$$g(b) = \sum_{\substack{i<j \\ \sigma(i)<\sigma(j)}} \frac{x_{\sigma(i)}}{x_{\sigma(i)} + x_{\sigma(j)}} + \sum_{\substack{i<j \\ \sigma(i)>\sigma(j)}} \left(1 + \frac{x_{\sigma(j)}}{x_{\sigma(i)} + x_{\sigma(j)}} \right).$$

Since

$$(-1)^{\mathrm{card}\{(i,j)|i<j \text{ and } \sigma(i)>\sigma(j)\}} = \epsilon(\sigma),$$

we obtain $g(b) = b$ if and only if $\epsilon(\sigma) = 1$, which is what we wanted.

Brief Solution to Exercise 9.3.7

Since n is coprime with the characteristic of k, the cardinality of $\mu_n(\bar{k})$ is n and P is separable over k. We know furthermore that it is a cyclic group: let us choose a generator ζ_n. Let $K = k(\zeta_n)$, $L = K(\sqrt[n]{a})$. Note that L does not depend on the choice of $\sqrt[n]{a}$. It is the splitting field of P. As P and $X^n - 1$ are separable over k, the fields L and K are Galois over k. We then have the fundamental exact sequence (Theorem 6.5.1, iv))

$$1 \to \mathrm{Gal}(L/K) \to \mathrm{Gal}(L/k) \to \mathrm{Gal}(K/k) \to 1.$$

But Gal(L/K) is cyclic thanks to Kummer's theory (Lemma 9.3.4) while Gal(K/k) is a subgroup of $(\mathbf{Z}/n\mathbf{Z})^*$ (Proposition 8.2.2), therefore it is abelian.

Brief Solution to Exercise 10.3.5

Let P be a monic annihilating polynomial with integer coefficients of z. We can assume $z \neq 0$. We already observe that all the z_i are non-zero since they are conjugates of z under the action of the Galois group. Then, the z_i, $i = 1, \ldots, d$ are roots of P as usual, and therefore are integers. Let G be the Galois group of $\mathbf{Q}(z_i)$ over $\mathbf{Q}$. As $\pi_n = \prod(X - z_i^n)$ is fixed by G, it has coefficients in $\mathbf{Q}$. But its coefficients are polynomials with integer coefficients in the z_i, therefore they are integers over $\mathbf{Z}$, and hence they are integers (Exercise 8.3.6). More precisely, these coefficients $a_j(n) \in \mathbf{Z}$ are elementary symmetric functions, a sum of $\binom{d}{j}$ products $z_{i_1}^n \cdots z_{i_j}^n$. We deduce the inequality $|a_j(n)| \leq \binom{d}{j}$: we therefore have a finite number of coefficients. There are therefore a finite number of polynomials π_n and hence a finite number of d-tuples (z_i^n). Let therefore $n < m$ such that $(z_i^n) = (z_i^m)$. Then $z_i^{m-n} = 1$.

Brief Solution to Exercise 10.7.2

See the solution to Exercise 12.9.

Solution to Exercise 12.1

(1) If $x = y^2$ with $y \in K$, we have $N_{K/\mathbf{Q}}(x) = [N_{K/\mathbf{Q}}(y)]^2$ and $N_{K/\mathbf{Q}}(y) \in \mathbf{Q}$.

   Now $N_{K/\mathbf{Q}}(4 + 2\sqrt{2}) = (4 + 2\sqrt{2})(4 - 2\sqrt{2}) = 8$ is not a square in $\mathbf{Q}$. Therefore, $4 + 2\sqrt{2}$ is not a square in K.

(2) We have $\mathbf{Q} \subset K \subset L = K[\sqrt{4 + 2\sqrt{2}}]$. According to the previous question, we have $[L : K] = 2$ and therefore $[L : \mathbf{Q}] = 4$

   The minimal polynomial over $\mathbf{Q}$ is $P(X) = \prod_{i=1}^{4}(X - x_i)$ where the x_i are the conjugates of $x_1 = \sqrt{4 + 2\sqrt{2}}$, that is

$$P(X) = \left(X + \sqrt{4 + 2\sqrt{2}}\right)\left(X + \sqrt{4 - 2\sqrt{2}}\right)\left(X - \sqrt{4 + 2\sqrt{2}}\right)$$
$$\times \left(X - \sqrt{4 - 2\sqrt{2}}\right)$$

   Similarly, the minimal polynomial over K is $\left(X - \sqrt{4 + 2\sqrt{2}}\right)$ $\left(X + \sqrt{4 + 2\sqrt{2}}\right)$.

(3) We have

$$\sqrt{4 + 2\sqrt{2}}\sqrt{4 - 2\sqrt{2}} = \sqrt{8} = 2\sqrt{2} \in K \subset L.$$

Therefore, $\sqrt{4 - 2\sqrt{2}} \in L$ as a quotient of elements of L.

Therefore all the conjugates of $\sqrt{4 + 2\sqrt{2}}$, namely $\pm\sqrt{4 \pm 2\sqrt{2}}$, are in L. As $\mathbf{Q}$ is perfect, $L/\mathbf{Q}$ is Galois.

(4) The Galois group $G = \mathrm{Gal}(L/K)$ is of order 4: it is therefore, up to isomorphism, $\mathbf{Z}/2\mathbf{Z} \times \mathbf{Z}/2\mathbf{Z}$ or $\mathbf{Z}/4\mathbf{Z}$. The action of G on the roots is transitive: let $\sigma \in G$ such that $\sigma(\sqrt{4 + 2\sqrt{2}}) = \sqrt{4 - 2\sqrt{2}}$. By squaring we get $\sigma(\sqrt{2}) = -\sqrt{2}$. But $\sqrt{4 + 2\sqrt{2}}\sqrt{4 - 2\sqrt{2}} = 2\sqrt{2}$ hence

$$\sigma\left(\sqrt{4 - 2\sqrt{2}}\right) = \frac{-2\sqrt{2}}{\sqrt{4 - 2\sqrt{2}}} = -\sqrt{4 + 2\sqrt{2}}.$$

Therefore $\sigma^2 = -\mathrm{Id}$ and σ of order 4: $G \simeq \mathbf{Z}/4\mathbf{Z}$.

(5) The subfields of L contain $\mathbf{Q}$ because the characteristic of L is 0. Therefore by the Galois correspondence, they bijectively correspond to the subgroups of $\mathbf{Z}/4\mathbf{Z}$: there are 3. The subfields are therefore $\mathbf{Q}$, K and L.

Solution to Exercise 12.2

(1) We can choose for d_i the i-th prime number.
(2) K_n is the splitting field of the polynomial $\prod_{i=1}^{n}(X^2 - d_i)$ which is separable over $\mathbf{Q}$.
(3) By the induction hypothesis

$$[K_n : \mathbf{Q}] = 2^n \text{ and } [K_{n-1} : \mathbf{Q}] = 2^{n-1}$$

and therefore $[K_n : K_{n-1}] = 2$. Thus $K_n = K_{n-1}(\sqrt{d_n})$ and the result follows.
(4) The image by σ of $\sqrt{d_{n+1}} \in K_n$ is a root of the polynomial $X^2 - d_{n+1} \in \mathbf{Q}[X]$.
(5) If $\epsilon = 1$, then $\sqrt{d_{n+1}}$ is invariant under $\mathrm{Gal}(K_n/K_{n-1})$. If $\epsilon = -1$, $\sqrt{d_{n+1}d_{n-1}}$ is invariant under $\mathrm{Gal}(K_n/K_{n-1})$. These elements are therefore in K_{n-1} (Galois' theorem). But this is absurd by the induction hypothesis applied to $(d_{n+1}, d_{n-1}, \ldots, d_1)$ or to $(d_{n+1}d_n, d_{n-1}, \ldots, d_1)$.

We can conclude the proof by induction (the base case is done for $n = 1$, which is a well-known case).
(6) The map is clearly an injective morphism, therefore bijective because the sets are finite of the same cardinality.
(7) As an abelian group, $\mathrm{Gal}(K_n/\mathbf{Q})$ has a unique $\mathbf{Z}$-module structure. As all elements are of order ≤ 2, this structure induces a unique structure of a $\mathbf{Z}/2\mathbf{Z}$-vector space. More explicitly, we simply set $n \cdot \sigma = \sigma^n$ for $n \in \mathbf{Z}/2\mathbf{Z}$ and $\sigma \in \mathrm{Gal}(K_n/\mathbf{Q})$. By cardinality, we obtain that the dimension is necessarily n.
(8) By the Galois correspondence, a subfield of degree 2 corresponds to a subgroup of $\mathrm{Gal}(K_n/\mathbf{Q})$ of cardinal 2^{n-1}, in other words to a hyperplane of the $\mathbf{Z}/2\mathbf{Z}$-vector space $(\mathbf{Z}/2\mathbf{Z})^n$. This corresponds to a non-zero linear form: there are exactly $2^n - 1$ of them.

Solution to Exercise 12.3

(1) Using the Chinese Lemma, we obtain the Galois group G

$$G \simeq (\mathbf{Z}/35\mathbf{Z})^* \simeq (\mathbf{Z}/5\mathbf{Z} \times \mathbf{Z}/7\mathbf{Z})^* \simeq (\mathbf{Z}/4\mathbf{Z} \times \mathbf{Z}/6\mathbf{Z}) \simeq \mathbf{Z}/12\mathbf{Z} \times \mathbf{Z}/2\mathbf{Z}.$$

This group has two quotients of order 2 (quotients by $\mathbf{Z}/12\mathbf{Z} \times 1$ and by $(2\mathbf{Z}/12\mathbf{Z}) \times \mathbf{Z}/2\mathbf{Z}$), so it is not cyclic.

(2) A subfield of degree 12 corresponds (Galois' theorem) to a subgroup of G of cardinality $24/12 = 2$. In other words, there are as many such subfields as there are elements of order 2 in G. We therefore solve $2(x, y) = (0, 0)$ in $\mathbf{Z}/12\mathbf{Z} \times \mathbf{Z}/2\mathbf{Z}$. We find 3 non-zero solutions

$$(6, 0), (6, 1), (0, 1).$$

So there are three subfields.

For fields of degree 6, we look for subgroups of order 4. They are either cyclic (a), or isomorphic to $\mathbf{Z}/2\mathbf{Z} \times \mathbf{Z}/2\mathbf{Z}$ (b). In case (a), we solve $4(x, y) = 0$ and $2(x, y) \neq 0$, which gives x of order 4 and y arbitrary. We therefore obtain two subgroups. In case (b), the non-zero elements are of order 2. We thus find a single subgroup

$$(6, 0), (6, 1), (0, 1), (0, 0).$$

In conclusion, we obtain 3 such subfields.

Solution to Exercise 12.4

(1) We have $\mathrm{disc}(P) = (-1)^{\frac{n(n-1)}{2}} \prod_{i \neq j}(z_i - z_j)$ for $P(X) = \Pi_i(X - z_i)$. Then

$$\prod_{k=1}^{n} P'(z_k) = \prod_{k=1}^{n} \prod_{i \neq k}(z_k - z_i) = \prod_{i \neq k}(z_k - z_i)$$

from which the formula follows. In the case $P(X) = X^q - 1$ we obtain

$$P'(z_k) = q z_k^{q-1} = q z_k^{-1}$$

$$\mathrm{disc}\,(P) = (-1)^{\frac{q(q-1)}{2}} q^q \prod_{k=1}^{q} z_k^{-1} = (-1)^{\frac{q(q-1)}{2}} q^q.$$

(2) We have $(X^q - 1)' = q X^{q-1}$ with root 0 in $\bar{k}$ because $q \neq 0$ in k. Therefore

$$\mathrm{GCD}(X^q - 1, (X^q - 1)') = 1$$

and $X^q - 1$ is separable. The rest follows directly from the course.

(3) The prime number q appears q times in the factorization of q^* into a product of prime factors. Since q is odd, q^* is not a square.

(4) The group $G = (\mathbf{Z}/q\mathbf{Z})^*$ is cyclic of order $q - 1$. However, $(g^{\frac{q-1}{2}})^2 = g^{q-1} = 1$ so $g^{\frac{q-1}{2}}$ is of order 1 or 2. It therefore belongs to the unique subgroup of order 2 of $(\mathbf{Z}/q\mathbf{Z})^* \simeq \mathbf{Z}/(q-1)\mathbf{Z}$, namely $\{-1, 1\}$ ($-1 \neq 1$ because q is odd).

(5) H is the kernel of a group morphism

$$
\begin{cases}
G \xrightarrow{\Psi} \pm 1 \\
g \longrightarrow g^{\frac{q-1}{2}}
\end{cases}.
$$

However, as $G \simeq \mathbf{Z}/(q-1)\mathbf{Z}$, Ψ is surjective so $|H| = \frac{|G|}{2} = \frac{q-1}{2}$. Since G is cyclic, it has a unique subgroup of index 2.

(6) We have

$$
\gamma(g)^{\frac{q-1}{2}} = g^{q-1} = 1 \text{ with } \mathrm{Im}(\gamma) \subset H.
$$

However, $\mathrm{Ker}(\gamma)$ consists of elements of order less than or equal to 2. Since G is cyclic there are 2 of them, namely ± 1 so $|\mathrm{Im}(\gamma)| = \frac{|G|}{|\mathrm{Ker}(\gamma)|} = \frac{q-1}{2}$ and $\mathrm{Im}(\gamma) = H$.

(7) ε is never trivial (otherwise G would be contained in A_q, see question (3)). So $|\mathrm{Ker}\varepsilon| = \frac{q-1}{2}$ and $\mathrm{Ker}\varepsilon = H$ We therefore have a factorization

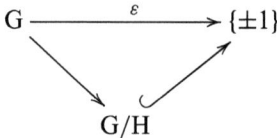

Since $|G/H| = 2$, $G/H \to \{\pm 1\}$ is *the unique* isomorphism of G/H onto $\{\pm 1\}$. Since we have a commutative diagram

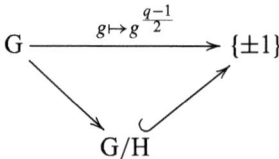

we deduce that $\varepsilon(g) = g^{\frac{q-1}{2}}$ for all $g \in G$.

(8) G acts transitively on the conjugates of ζ, which are the primitive q^{th} roots of 1. Since $p \neq q$, ζ^p is a primitive root. Hence the existence of Φ. Since ζ generates the group of qth roots of 1, Φ is unique.

(9) The image of Φ is p mod $q \in (\mathbf{Z}/q\mathbf{Z})^*$, so $\Phi \in H$ if and only if $p^{\frac{q-1}{2}} = 1(q)$. However, according to the course, $\Phi(\sqrt{q^*}) = \varepsilon(\Phi)\sqrt{q^*}$ so $\Phi \in H$ if and only if $\varepsilon(\Phi) = 1$, that is $p^{\frac{q-1}{2}} = 1$. We get the answer after (4).

(10) This follows directly from the course.

(11) Set $A = \mathbf{Z}[\zeta]$. We have a commutative diagram

$$
\begin{array}{ccc}
\mathbf{Z}[\zeta] = A & \longrightarrow & A/pA \\
\downarrow{\scriptstyle \Phi} & & \\
& \longrightarrow & A/p \\
\downarrow{\scriptstyle \Phi mod p} & & \downarrow{\scriptstyle \tilde{\Phi}} \\
\mathbf{Z}[\zeta] = A & \longrightarrow A/pA & \longrightarrow A/p
\end{array}
$$

Then $\Phi(P(\zeta)) = P(\xi^p)$ mod pA so Φ mod $p = F_p$ and $\tilde{\Phi}$ is the Frobenius morphism.

(12) We can write the commutative diagram

$$
\begin{array}{ccc}
F_p \in Gal(F/Fp) & \lhook\joinrel\longrightarrow & Sym(\overline{(z_i)}) \\
\uparrow{\scriptstyle Z} & & \uparrow{\scriptstyle Z} \\
\Phi \in D_p & \lhook\joinrel\longrightarrow & Sym(z_i)
\end{array}
$$

which gives the result.

(13) According to the previous question, we have $\Phi \in H$ if F_p is in the alternating group, that is if q^* mod p is a square in $\mathbf{F}_p$, or even if

$$
\left((-1)^{\frac{q-1}{2}} q\right)^{\frac{p-1}{2}} = (-1)^{\frac{q-1}{2}\frac{p-1}{2}} \left(\frac{q}{p}\right).
$$

(14) It is enough to use the formula from the previous question.

Solution to Exercise 12.5

(1) Working with $i \in \mathbf{Z}/m\mathbf{Z}$, we write

$$
(\sigma(a_1, \ldots, a_m)\sigma^{-1})\sigma(a_i) = (\sigma(a_1, \ldots, a_m))a_i = \sigma(a_{i+1}).
$$

(2) According to the previous question, $s^i \sigma s^{-i}$ fixes $\beta = s^i(\alpha)$, where α is the point such that $\sigma(\alpha) = \alpha$. As s is an n-cycle of S_n, it acts transitively on $\{1, \ldots, n\}$. We therefore choose i such that $s^i(\alpha) = b$.

(3) If necessary, by changing σ to $s^i \sigma s^{-i}$, we can assume $\sigma(a) = a$. We have in $\sigma^i(a, b)\sigma^{-1} = (a, \sigma^i(b))$ according to the first question. But σ is an $(n-1)$

cycle fixing a and therefore acts transitively on $\{1, ..., n\} - \{a\}$. We can thus choose i such that $\sigma^i(b) = a + 1$. We then have $\sigma^i(a, b)\sigma^{-i} = (a, a+1) \in \Sigma$.

We then have $s^j(a, a + 1)s^{-j} = (a + j, a + j + 1)$ for $j \in \mathbf{Z}/n\mathbf{Z}$, which gives the result.

(4) It is enough to notice that the set

$$\{(i, (i + 1)), 1 \leq i < n\}$$

generates S_n.

(5) We can consider the minimal polynomial of a primitive element of $\mathbf{F}_{p^n}/\mathbf{F}_p$.

(6) Let p be a prime number $p \geq \sup(n - 2, 5)$ (there certainly exists one). According to the previous question, we can choose

$$P_2, P_3 \text{ and } P_p \text{ respectively in } \mathbf{F}_2[X], \mathbf{F}_3[X] \text{ and } \mathbf{F}_p[X]$$

monic irreducible (these are the candidates to be the reductions for the sought polynomial). As $\mathrm{GCD}(2, 3, p) = 1$, the Chinese Remainder Theorem ensures the existence of $P \in \mathbf{Z}[X]$ such that

$$P \equiv P_2 \bmod (2),$$
$$P \equiv P_3 \bmod (3),$$
$$P \equiv P_p \bmod (p).$$

The dominant coefficient $a_n(P)$ of P satisfies

$$a_n(P) \equiv 1 \bmod (2 \cdot 3 \cdot p),$$

that is to say $a_n(P) = 1 + 2 \cdot 3pk$ with $k \in \mathbf{Z}$. By replacing P with $P - 2 \cdot 3pkX^n$, we obtain the result.

(7) It is enough to use question (4) and the reduction theorem mod 2, 3 and p.

Solution to Exercise 12.6

(1) We have $\mathbf{Q}[\zeta] = \mathbf{Q}[\zeta + \zeta^{-1}][\mathrm{Im}(\zeta)\mathrm{I}]$. However, $(\mathrm{Re}(\zeta))^2 - (\mathrm{Im}(\zeta)\mathrm{I})^2 = 1$ and, as $n \geq 3$, $\zeta \notin \mathbf{R}$. Therefore $[\mathbf{Q}[\zeta] : \mathbf{Q}[\zeta + \zeta^{-1}]] = 2$. We obtain the result by using the telescopic base theorem.

(2) We have $\mathbf{Q}[\zeta] \supsetneq \mathbf{R} \cap \mathbf{Q}[\zeta] \supset \mathbf{Q}[\zeta + \zeta^{-1}]$ and the result follows by reasons of dimension.

(3) As $\bar{\zeta} = \zeta^{-1}$, the image is -1.

(4) The group $\mathrm{Gal}(\mathbf{Q}[\zeta] : \mathbf{Q})$ being abelian, every subgroup is normal and therefore every subfield is Galois over $\mathbf{Q}$. Now

$$\mathrm{Gal}(\mathbf{Q}[\zeta] : \mathbf{Q}[\zeta] \cap \mathbf{R}) = \{\mathrm{Id}, \text{ conjugation}\}.$$

Therefore $\mathrm{Gal}(\mathbf{Q}[\zeta + \zeta^{-1}], \mathbf{Q}) = (\mathbf{Z}/n\mathbf{Z})^*/\{\pm 1\}$. Note that as $n \geq 3$, we have $-1 \neq 1$.

(5) Let us write $x = \frac{a}{b}$ with $b > 0$ and $gcd(a, b) = 1$.

If $b \geq 3$ the degree is $\varphi(b)/2$ according to question 1.

If $b = 1$ or $b = 2$ the degree is clearly 1.

(6) We start by expressing $\sin(\frac{2\pi}{n})$ in terms of $\cos(\frac{2\pi}{n})$ using the trigonometric formula $\sin(\frac{2\pi}{n}) = \cos(\pi/2 - \frac{2\pi}{n})$. We can then discuss the different cases using question 5.

(7) It is a determinant divided by 2.

(8) We consider such a regular polygon with n sides and vertices in $\mathbf{Q}^2$. The centroid O is then in $\mathbf{Q}^2$. An elementary triangle with vertices O, M, M$'$ where M, M$'$ are two consecutive vertices of the polygon has an area of $\frac{OM^2}{2} \sin \frac{2\pi}{n}$.

Therefore, according to the previous question, we obtain $\sin\left(\frac{2\pi}{n}\right) \in \mathbf{Q}$. We can then apply question 6, which gives $n = 4$.

(9) According to the course, we first have $\mathbf{Q}[I] \cap \mathbf{Q}[\exp(\frac{2I\pi}{n})] = \mathbf{Q}[\exp(\frac{2I\pi}{d})]$ with $d = gcd(4, n)$. Thus $I \in \mathbf{Q}(\zeta)$ is equivalent to $\mathbf{Q}[\exp(\frac{2I\pi}{d})] = \mathbf{Q}(I)$, that is to say $d = 4$.

(10) We look at the extensions

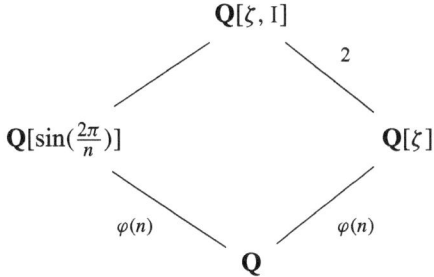

So we have $[\mathbf{Q}[\zeta, I] : \mathbf{Q}[\sin(\frac{2\pi}{n})]] = 2$. Since I is not real and is not in $\mathbf{Q}[\sin(\frac{2\pi}{n})]$, we have $\mathbf{Q}[\zeta, I] = \mathbf{Q}[\sin(\frac{2\pi}{n}), I]$.

However, $[\mathbf{Q}[\sin(\frac{2\pi}{n}), I] : \mathbf{Q}[\sin(\frac{2\pi}{n})]] \leq 2$ and $[\mathbf{Q}[I\sin(\frac{2\pi}{n})] : \mathbf{Q}] \leq \varphi(n)$ because $\mathbf{Q}[I\sin(\frac{2\pi}{n})] \subset \mathbf{Q}[\zeta]$. We deduce that $[\mathbf{Q}[I\sin(\frac{2\pi}{n})] : \mathbf{Q}] = \varphi(n)$.

(11) We have $\sin\left(\frac{2\pi}{n}\right) \in \mathbf{Q}[\zeta] \cap \mathbf{R} = \mathbf{Q}[\cos\left(\frac{2\pi}{n}\right)]$ according to question 2.

For degree reasons, we have $\mathbf{Q}[\sin\left(\frac{2\pi}{n}\right)] = \mathbf{Q}[\cos\left(\frac{2\pi}{n}\right)]$

(12) We can assume $\zeta = \exp\frac{2I\pi}{n}$, which is indeed primitive. Then

$$-2\sin\left(\frac{2\pi}{n}\right) = \zeta^{n/4}(\zeta - \zeta^{-1}).$$

If $g_m, m \in (\mathbf{Z}/n\mathbf{Z})^*$ is the element of $\mathrm{Gal}(\mathbf{Q}[\zeta], \mathbf{Q})$ corresponding to m, we have

$$g_m\left(-2\sin\left(\frac{2\pi}{n}\right)\right) = [\zeta^{n/4}]^m \zeta^m - \zeta^{-m} = \mathrm{I}^m 2\mathrm{I}\sin\left(\frac{2m\pi}{n}\right).$$

So

$$g_m\left(\sin\frac{2\pi}{n}\right) = -\mathrm{I}^{(m+1)}\sin\left(\frac{2m\pi}{n}\right).$$

Since $gcd(m, n) = 1$, we obtain $\varphi(n)/2$ distinct conjugates.

Solution to Exercise 12.7

(1) This is a result from the course.
(2) We identify A as a sub-ring of B. Let b be the inverse of $a \neq 0$ in B (B is a field). It is enough to show that $b \in$ A. We have an equality

$$b^n + a_{n-1}b^{n-1} + \cdots + a_0 = 0,$$

with the $a_i \in$ B. Multiply by a^n:

$$1 = -(a_{n-1}a + \cdots + a^n a_0)$$

$$= a(-a^{n-1} \cdots - a_{n-1}a_0).$$

So $b = (-a^{n-1} - \cdots - a_{n-1}a_0)$ and the result follows.
(3) This is an immediate consequence of the previous question.
(4) This is a result from the course.
(5) By definition, k-algebra morphism $k[\mathrm{T}] \rightarrow k[x_n]$ such that $\mathrm{T} \mapsto x_n$ is bijective. Since B is a field, the inclusion $k[x_n] \xrightarrow{\varphi} \mathrm{B}$ extends to the field of fractions by setting $\varphi\left(\frac{\mathrm{P}(x_n)}{\mathrm{Q}(x_n)}\right) = \frac{\varphi(\mathrm{P}(x_n))}{\varphi(\mathrm{Q}(x_n))}$.
(6) We have

$$\mathrm{B} = k[x_1, \ldots, x_n] = k(x_n)[x_1, \ldots, x_{n-1}].$$

We apply the induction hypothesis to the K-algebra B where

$$\mathrm{K} = k(x_n).$$

(7) The x_i, $i = 1, \ldots, n-1$, are integers over $k[x_n]$. We take P as a common denominator of the coefficients of the equations. As the elements of B that are integers over $k[x_n, \mathrm{P}^{-1}]$ form a subring, B is integer over $k[x_n, \mathrm{P}^{-1}]$.

(8) Let N be the maximum of the degrees of the minimal polynomials of the x_i in $k[x_n, P^{-1}][T]$. Then B is generated by the monomials $x_1^{\alpha_1} \cdots x_{n-1}^{\alpha_{n-1}}$ with $\alpha_i \leq N$.

   We then apply the result of question 3.

(9) This is immediate.

(10) $k[x_n, P^{-1}]$ is a field so its ideals are trivial.

(11) It suffices to proceed by induction on n. The result is clear for $n = 0$, and we have shown the implication $(n \Rightarrow n + 1)$.

(12) If I is a maximal ideal, $k[x_1, \ldots, x_n]/I$ is of finite dimension. As k is algebraically closed, the morphism $k \to k[X_1, \ldots, X_n]/I$ is bijective. Therefore, the morphism

$$k[X_1, \ldots, X_n]/I \leftarrow k[X_1, \ldots, X_n]$$

is identified with a morphism of algebras

$$\begin{cases} k[X_1, \ldots, X_n] \xrightarrow{\varphi} k \\ X_i \qquad\qquad \mapsto x_i \end{cases}$$

So we have $I = \mathrm{Ker}\varphi = I_x$. The converse is immediate. The last assertion follows from the construction of I.

Solution to Exercise 12.8

(1) We write x as proposed. Then $n(\deg(P) - \deg(Q)) = 1$, a contradiction with $n > 1$.

(2) The conjugates of $\sqrt[n]{t}$ are of the form $\sqrt[n]{(t)} \exp(\frac{2lk\pi}{n}) \in K_n$, so the extension is Galois. It is a cyclic extension. As t has no d-th root in k for $d \geq 2$, the Galois group is isomorphic to $\mathbf{Z}/(n\mathbf{Z})$ according to the course.

(3) We can use the Galois correspondence, and we get as many subfields as subgroups of $\mathbf{Z}/(n\mathbf{Z})$, that is to say as many integers d dividing n. These are the $K[(\sqrt[n]{t})^d] = K_{n/d}$.

(4) As K_{nm}/K is Galois and $K \subset K_n \subset K_{nm}$, the extension K_{nm}/K_n is also Galois. According to the course, H_n is a subgroup of H.

(5) We have $K \subset K_n K_m \subset K_{nm}$ so $K_{nm}/K_n K_m$ is Galois. Let G be its Galois group. Then G is formed of the elements of H leaving K_n and K_m fixed. It is therefore isomorphic to $H/(H_n H_m)$ where $H_n H_m$ is the subgroup of H generated by H_n and H_m.

(6) According to the previous question, we have $[K_n K_m : K] = [K_{\varpi} : K]$. According to question 3, this implies the equality $K_n K_m = K_{\varpi}$.

(7) As above, the extension is clearly Galois, and its Galois group is isomorphic to $H/(H_n \cap H_m)$.

(8) We have equality as above.

Solution to Exercise 12.9

(1) The inclusion $K \subset \mathbf{Q}(\zeta, \sqrt[\ell]{n})$ is clear because the roots of P are the $\sqrt[\ell]{n}\zeta^r$, $r \in \mathbf{Z}$. For the other inclusion, it is enough to notice that $\zeta = (\zeta \sqrt[\ell]{n})/\sqrt[\ell]{n}$.

(2) If $x = 0$ it is clear. Suppose $x \neq 0$. As ℓ is prime, we have $\alpha, \beta \in \mathbf{Z}$ such that $\alpha(\ell - 1) + \beta\ell = 1$. Then $x = x^{\beta\ell}y^{\ell\alpha}$ is an ℓ-power in $\mathbf{Q}$. If x is an integer, write $y = p/q$ in irreducible form. Then $p^{\ell} = x^{\ell-1}q^{\ell}$ so $q = \pm 1$ and $y = p$ is an integer, which implies the result.

(3) As $[\mathbf{Q}[\zeta] : \mathbf{Q}] = \ell - 1$, we have $N_{\mathbf{Q}[\zeta]/\mathbf{Q}}(n) = n^{\ell-1}$. If $\sqrt[\ell]{n} \in \mathbf{Q}[\zeta]$, we also have $N_{\mathbf{Q}[\zeta]/\mathbf{Q}}(n) = (N_{\mathbf{Q}[\zeta]/\mathbf{Q}}(\sqrt[\ell]{n}))^{\ell}$. So $n^{\ell-1} = (N_{\mathbf{Q}[\zeta]/\mathbf{Q}}(\sqrt[\ell]{n}))^{\ell}$. According to the previous question, n is an ℓ-power in $\mathbf{Z}$, a contradiction. Therefore $n \notin \sqrt[\ell]{n} \in \mathbf{Q}[\zeta]$ and n is not an ℓ-power in $\mathbf{Q}[\zeta]$.

(4) The extension $K/\mathbf{Q}[\zeta]$ is cyclic. The only divisor of ℓ different from 1 is ℓ, and n is not an ℓ-power in $\mathbf{Q}[\zeta]$ according to the previous question. Therefore, according to the course, P is irreducible over $\mathbf{Q}[\zeta]$.

(5) Still according to the course, $\mathrm{Gal}(K/\mathbf{Q}[\zeta]) \simeq \mathbf{Z}/(\ell\mathbf{Z})$. The Galois correspondence gives an exact sequence:

$$0 \to \mathbf{Z}/(\ell\mathbf{Z}) \to G \to \mathrm{Gal}(\mathbf{Q}[\zeta]/\mathbf{Q}) \to 1.$$

According to the course, for the cyclotomic extension (ℓ is prime):

$$\mathrm{Gal}(\mathbf{Q}[\zeta]/\mathbf{Q}) \simeq (\mathbf{Z}/\ell\mathbf{Z})^* \simeq \mathbf{Z}/((\ell - 1)\mathbf{Z}).$$

Solution to Exercise 12.10

(1) If $z \neq 1$, the matrix is diagonalizable, so $M = D$ and $N = 0$. If $z = 1$, we have $D = I_2$ and $N = M - I_2$.

(2) In the studied example, the functions are clearly not continuous at 1. The general functions $M \mapsto D$ and $M \mapsto N$ are therefore not continuous.

(3) It is enough to regard M, N as matrices of $M_n(\overline{\mathbf{Q}})$ and apply the theorem.

(4) It is enough to consider the extension K of $\mathbf{Q}$ generated by the coefficients of M and N and their conjugates over $\mathbf{Q}$. As these coefficients are algebraic over $\mathbf{Q}$, K is finite. It is Galois by construction.

(5) The action of G on K extends to an action on $M_n(K)$. Then for $g \in G$, we have $g(M) = g(D) + g(N)$. It is easy to see that this is the Jordan–Chevalley decomposition of $g(M) = M$. Therefore $g(D) = D$ and $g(N) = N$. This being true for all $g \in G$, we have $D, N \in M_n(\mathbf{Q})$.

(6) The proof is the same.

(7) Let $\sqrt{t}$ be a root of t in $\bar{k}$. Then the decomposition is $D = \sqrt{t}I_2$ and $N = M - \sqrt{t}I_2$, which are not coefficients in k.

Solution to Exercise 12.11

(1) We have $\mathrm{disc}(P) = -(z_1 - z_2)(z_1 - z_3)(z_2 - z_1)(z_2 - z_3)(z_3 - z_1)(z_3 - z_2)$. If two roots are the same, the two terms are zero. Otherwise, we have $P'/P = (X - z_1)^{-1} + (X - z_2)^{-1} + (X - z_3)^{-1}$ and therefore $P'(z_i) = (z_i - z_j)(z_i - z_k)$ for i, j, k distinct. This implies the formula.

Now, $P'(X) = 3X^2 + p$, $z_1z_2z_3 = -q$, $z_1 + z_2 + z_3 = 0$, $z_1z_2 + z_2z_3 + z_1z_3 = p$.
So $\mathrm{disc}(P) = -p^3 - 27q^2 - 9p(z_1^2z_2^2 + z_1^2z_3^2 + z_2^2z_3^2) - 3p^2(z_1^2 + z_2^2 + z_3^2)$.
We have $z_1^2 + z_2^2 + z_3^2 = -2(z_1z_2 + z_1z_3 + z_2z_3) = -2p$.
But $z_1^2z_2^2 + z_1^2z_3^2 + z_2^2z_3^2 = q^2(z_1^{-2} + z_2^{-2} + z_3^{-2})$. By summing the $z_i + pz_i^{-1} + qz_i^{-2} = 0$, we get $q(z_1^{-2} + z_2^{-2} + z_3^{-2}) = -p(z_1^{-1} + z_2^{-1} + z_3^{-1}) = pq^{-1}(z_1z_2 + z_1z_3 + z_2z_3) = p^2q^{-1}$.
So $z_1^2z_2^2 + z_1^2z_3^2 + z_2^2z_3^2 = p^2$.
In the end $\mathrm{disc}(P) = -p^3 - 27q^2 - 9p^3 + 6p^3 = -4p^3 - 27q^2$.

(2) If there is a root in $\mathbf{Q}$, it is reducible. If it is reducible, it has at least one factor of degree 1.
Suppose P is reducible. If P has its 3 roots rational, then G is the trivial group. If P has two rational roots and one non-rational root, an element of G can only act trivially on the set of roots, so G is also trivial. Suppose that P has a single rational root z_1. Then $P = (X - z_1)Q$ with Q rational irreducible. So G is the Galois group of Q and therefore G is of order 2 isomorphic to $\mathbf{Z}/2\mathbf{Z}$.

(3) If P has a double root, it is fixed by all elements of G, so it is rational, a contradiction.

(4) S_3 is of order 6. We have the trivial subgroups of order 1 and 6. The other subgroups are of order 2 and 3, so cyclic. The subgroups of order 2 are generated by the transpositions (there are 3 of them). The subgroup of order 3 is generated by a 3-cycle (it is A_3).

(5) If $\mathrm{disc}(P)$ is a square, then $G \subset A_3$. As G is not trivial, we then have $G = A_3$. If $\mathrm{disc}(P)$ is not a square, then G is not included in A_3. But G is not of order 2 because it acts transitively on the roots (P is irreducible). Therefore, $G = S_3$.

(6) If $\mathrm{disc}(P)$ is a square, then $G = A_3$ only has trivial subgroups, so the subfields of K are $\mathbf{Q}$ and K. Otherwise, $G = S_3$ and K has 6 subfields. We have the trivial subfields K and $\mathbf{Q}$. We have a subfield of degree 2 over $\mathbf{Q}$, namely $\mathbf{Q}[\sqrt{\mathrm{disc}(P)}]$. We have 3 subfields of degree 3, namely $\mathbf{Q}[z_1], \mathbf{Q}[z_2], \mathbf{Q}[z_3]$.

(7) If $\mathrm{disc}(P)$ is a square, then $\mathbf{Q}(\sqrt{\mathrm{disc}(P)}) = \mathbf{Q}$ and it is clear. Otherwise, for degree reasons we have $K = [\mathbf{Q}(\sqrt{\mathrm{disc}(P)})][z_1]$. Therefore, P is the minimal polynomial of z_1 over $\mathbf{Q}(\sqrt{\mathrm{disc}(P)})$ and is therefore irreducible over this field.

Solution to Exercise 12.12

(1) G_d is the set of solutions x in $\mathbf{Z}/n\mathbf{Z}$ of the equation $dx = 0$. We have exactly d solutions. We obtain the subgroup $G_d = ((n/d)\mathbf{Z})/n\mathbf{Z}$.

(2) G_d is normal in a commutative group. The quotient is clearly cyclic generated by the image of 1. It is of cardinality n/d.

(3) This is the theorem of the primitive element. The polynomial P is monic, irreducible over $\mathbf{Q}$ and of degree n.

(4) If n is even, G has a subgroup $G_{n/2}$ of order $n/2$ (resp. G_2 of order 2) and therefore K has a subfield of degree 2 (resp. $n/2$) over $\mathbf{Q}$ according to the Galois correspondence. If n is odd, according to Lagrange's theorem, G does not have a subgroup of order 2 (or $n/2$ which is not an integer) and we can conclude in the same way.

(5) As G is commutative, K/L and L/$\mathbf{Q}$ are Galois. In the case $[K : L] = 2$, we have $\mathrm{Gal}(K/L) = G_2$ and $\mathrm{Gal}(L/\mathbf{Q}) = G/G_2$. In the case $[L : \mathbf{Q}] = 2$, we have $\mathrm{Gal}(K/L) = G_{n/2}$ and $\mathrm{Gal}(L/\mathbf{Q}) = G/G_{n/2}$.

(6) We have $P(\sigma(x)) = 0$ because P has coefficients in $\mathbf{Q}$, so $\sigma(x)$ is a conjugate of x and generates K. Therefore, $\sigma(K) = K$.

(7) The image of G in S_n is generated by an element of order n which acts transitively on $\{1, \ldots, n\}$, so it is an n-cycle. As n is odd, it is of signature 1. Therefore, G is included in A_n and the discriminant of P is a square.

(8) σ is of order 2 and G of odd order does not contain an element of order 2, hence the result.

(9) As K is not included in $\mathbf{R}$, σ is of order 2 in G and L' is a subfield of K of degree 2 over $\mathbf{Q}$. Therefore, $L = L'$ and $L \subset \mathbf{R}$.

(10) Suppose $\sqrt{m} \in K$. If $\sqrt{m} \in \mathbf{Q}$, we indeed have $m \geq 0$. Otherwise, we have $\mathbf{Q}[\sqrt{m}]$ as a subfield of K of degree 2 over $\mathbf{Q}$, so $\mathbf{Q}[\sqrt{m}] = L \subset \mathbf{R}$. Therefore m, being a square of a real number, is positive.

Solution to Exercise 12.13

(1) Since p does not divide n, the roots of $(X^n - 1)' = nX^{n-1}$ are 0. 0 not being a root of $X^n - 1$, we obtain the result.

(2) $\mu_n(K)$ is made up of powers of ζ, so $\mu_n(K) \subset A$. The map is therefore well defined. The image is indeed in $\mu_n(\kappa)$. As the elements of $\mu_n(K)$ are invertible in A, $\mu_n(K) \cap \mathfrak{p}$ is empty. Therefore, the map is injective and defines an isomorphism. The two groups are therefore cyclic (we know that $\mu_n(K)$ is). The image of ζ generates the image.

(3) The conjugates of ζ are powers of ζ, and therefore are in K. Therefore K/$\mathbf{Q}$ is Galois. As $\kappa = \mathbf{F}_p[\bar{\zeta}]$, the same applies. An element $g \in G$ is determined by $g(\zeta) = \zeta^r$ with $r \in (\mathbf{Z}/n\mathbf{Z})^*$, which gives the embedding. The argument for $\overline{G}$ is analogous.

(4) An element of $D(\mathfrak{p})$ is well defined on κ, and therefore we have the natural inclusion in G. The power of ζ and $\bar{\zeta}$ is the same for the two morphisms, hence the identification.

(5) The map $D(\mathfrak{p}) \to \overline{G}$ is surjective, hence the existence of $F_\mathfrak{p}$ (it is unique since the map is in fact an isomorphism). Then $F_\mathfrak{p}(\bar{\zeta}) = \bar{\zeta}^p$ and therefore the image of $F_\mathfrak{p}$ in G sends ζ to ζ^p, hence the result.

(6) We have shown that every number coprime with n is in G. Therefore $G = (\mathbf{Z}/n\mathbf{Z})^*$.

(7) From the above, $[K : \mathbf{Q}] = \phi(n)$. As this is the degree of Ψ_n, which is rational, Ψ_n is the minimal polynomial of ζ over $\mathbf{Q}$. Hence the result.

Solution to Exercise 12.14

(1) Let us first note that $\mathbf{Q}[\sqrt{21}] = \mathbf{Q} \oplus \mathbf{Q}\sqrt{21}$. Suppose we have $a, b \in \mathbf{Q}$ with

$$5 + \sqrt{21} = (a + b\sqrt{21})^2 = a^2 + 2ab\sqrt{21} + 21b^2.$$

Then $2ab = 1$ and $5 = a^2 + 21b^2 = a^2 + \frac{21}{4a^2}$. Therefore $a^4 - 5a^2 + 21/4 = 0$. So $a^2 = (5 \pm 2)/2$ and a cannot be in $\mathbf{Q}$, a contradiction.

(2) As $(x^2 - 5)^2 - 21 = 0$, we have $[K : \mathbf{Q}] \leq 4$. But K contains $\mathbf{Q}[\sqrt{21}]$ so 2 divides $[K : \mathbf{Q}]$. If $[K : \mathbf{Q}] = 2$, then $K = \mathbf{Q}[\sqrt{21}]$. Therefore $5 + \sqrt{21} = z^2$ is a square in $\mathbf{Q}[\sqrt{21}]$, a contradiction with the previous question.

(3) We have $zz' = \sqrt{25 - 21} = 2$, so $z' = 2z^{-1} \in K$. We have $\pi_{x,\mathbf{Q}}$ of degree 4, so

$$\pi_{x,\mathbf{Q}}(X) = (X^2 - 5)^2 - 21.$$

So the conjugates of x over $\mathbf{Q}$ are $\{z, -z, z', -z'\}$. They are all in K, so the extension is Galois.

(4) and (5) are deduced from the transitivity of the Galois group action on the conjugates and the fact that $K = \mathbf{Q}[z]$, which implies that an element of G is determined by its action on z.

(6) We have

$$g(z') = g(2z^{-1}) = -2z^{-1} = -z'$$

and

$$h(z') = h(2z^{-1}) = 2(z')^{-1} = z^{-1}.$$

Therefore

$$g(h(z)) = g(z') = -z' \text{ and } h(g(z)) = h(-z) = -z'.$$

Therefore $g \circ h = h \circ g$. As $K/\mathbf{Q}$ is Galois, we have $|G| = [K : \mathbf{Q}] = 4$. Furthermore

$$g^2(z) = g(-z) = z \text{ and } h^2(z) = h(z') = z.$$

Therefore $h^2 = g^2 = 1$ and the map

$$\phi : (\mathbf{Z}/2\mathbf{Z})^2 \to G, \phi(a, b) = h^a g^b$$

is well defined. As $hg = gh$, ϕ is a group morphism. This morphism is clearly injective, so it is an isomorphism.

(7) We first have $\mathbf{Q}$ and $K = \mathbf{Q}[x]$. It remains to determine the subfields of degree 2 over $\mathbf{Q}$. We apply the Galois theorem. G has 3 elements of order 2: g, h and gh. So we have 3 subfields of degree 2. These are

$$K^g = \mathbf{Q}[\sqrt{21}] = \mathbf{Q}[z^2] = \mathbf{Q}[(z')^2],$$

$$K^h = \mathbf{Q}[z + z'] \text{ and } K^{gh} = \mathbf{Q}[z - z'].$$

(8) We have $z = \sqrt{\frac{3}{2}} + \sqrt{\frac{7}{2}}$.

Solution to Exercise 12.15

(1) For each i, we have $P'(x_i) = \prod_{j \neq i}(x_i - x_j)$. Therefore

$$\mathrm{disc}(P) = (-1)^{\frac{n(n-1)}{2}} \prod_{i \neq j}(x_i - x_j) = (-1)^{\frac{n(n-1)}{2}} \prod_{1 \leq i \leq n} P'(x_i).$$

But $P'(X) = n \prod_{1 \leq j \leq n-1}(X - y_j)$, so

$$\mathrm{disc}(P) = n^n(-1)^{\frac{n(n-1)}{2}} \prod_{1 \leq i \leq n, 1 \leq j \leq n-1} (x_i - y_j)$$

$$= n^n(-1)^{\frac{n(n-1)}{2}+n(n-1)} \prod_{1 \leq i \leq n, 1 \leq j \leq n-1} (y_j - x_i)$$

$$= n^n(-1)^{\frac{n(n-1)}{2}+n(n-1)} \prod_{1 \leq j \leq n-1} P(y_j).$$

(2) Let $P(X) = X^n + aX + b$. Then $P'(X) = nX^{n-1} + a$ and P' has $n - 1$ roots $y_j = \xi^j \epsilon$ with ϵ root $(n - 1)^{\mathrm{th}}$ of $-a/n$ and ξ primitive root $(n - 1)^{\mathrm{th}}$ of 1. Then

$$\prod_{1 \leq j \leq n-1} P(y_j) = \prod_{1 \leq j \leq n-1} (-a\xi^j \epsilon/n + a\epsilon\xi^j + b)$$

$$= (-b)^{n-1} \prod_{1 \leq j \leq n-1} (-1 - ab^{-1}\epsilon\xi^j(n-1)/n)$$

$$= (-b)^{n-1}(X^{n-1} + a^n b^{1-n}(n-1)^{n-1}/n^n)(-1)$$

$$= b^{n-1} + a^n(1-n)^{n-1}/n^n.$$

Therefore

$$\text{disc}(P) = (-1)^{\frac{n(n-1)}{2}} (b^{n-1}n^n + a^n(1-n)^{n-1}).$$

(3) We have $P'(X) = 5X^4 + 20$. So if x is a root of P and of P', we have

$$x^4 = -4 \text{ and } 0 = -4x + 20x + 16.$$

So $x = 1$, a contradiction. Therefore $P \wedge P' = 1$ and P is separable.

(4) As P is of odd degree, P has at least one real root. But P' is strictly positive on **R**, so P has only one real root. The other roots of P are of the form $z_1, \overline{z_1}, z_2, \overline{z_2}$. The image of the complex conjugation in S_5 is therefore a double transposition $(z_1, \overline{z_1})(z_2, \overline{z_2})$.

(5) We have

$$\text{disc}(P) = (-1)^{10}((-4)^4 20^5 + 5^5 16^4) = 5^5 16^4 (4+1) = (16^2 5^3)^2 \in (\mathbf{Q}^*)^2.$$

So G is contained in S_5.

(6) Let $\tilde{P}$ be the reduction of P modulo 7. Then

$$\tilde{P}(X) = X^5 - X + 2 = (X+2)(X+3)(X^3 + 2X^2 - 2X + 5).$$

Since $X^3 + 2X^2 - 2X + 5$ of degree 3 has no root in $\mathbf{F}_7$, it is irreducible and separable. The theorem of reduction modulo 7 applies and we obtain a 3-cycle in G.

(7) We have $\overline{P}(X) = X^5 - X + 1$. For $x \in \mathbf{F}_9^*$, we have $x^8 = 1$, so $(x^4)^2 = 1$ and $x^4 = \pm 1$, so $x^5 = \pm x$. If x is also a root of $\overline{P}$; we have $0 = \pm x - x + 1$. So $2x = 1$ and $x = 2$, which is not a root of $\overline{P}$. Since 0 is not a root of $\overline{P}$, there are no roots in $\mathbf{F}_9$.

(8) If $\overline{P}$ was reducible, it would have a factor S of degree ≤ 2. Then we would have an extension $\mathbf{F}_3[X]/(S)$ in which $\overline{P}$ would have a root, so $\overline{P}$ would have a root in $\mathbf{F}_9$, a contradiction. So $\overline{P}$ is irreducible and separable. The reduction modulo 3 applies and so G contains at least one 5-cycle.

(9) Let G be a subgroup of S_4 generated by a 3-cycle σ and a double transposition $\tau = \tau_1 \tau_2$. We first have $G \subset A_4$. By conjugating by the 3-cycle, we obtain all the double transpositions in G. By conjugating the 3-cycle and therefore square by the double transpositions, we obtain all the 3-cycles in G. So $G = A_4$.

(10) Let G be a subgroup of S_5 generated by a 5-cycle, a 3-cycle and a double transposition. We first have $G \subset A_5$. By conjugating with the 3-cycle by the double transposition, then by the 5-cycle, we can assume that the support of the 3-cycle and the double transposition are included in $\{1, 2, 3, 4\}$. So we have $A_4 \subset G$. But $|A_5|/|A_4| = 5$ prime. So $G = A_4$ or $G = A_5$. Since G contains a 5-cycle, we have $G = A_5$.

(11) This is a direct consequence of the previous questions.

Solution to Exercise 12.16

(1) We have $G \simeq \mathbf{Z}/n\mathbf{Z}$, where n is the cardinality of G. Then a subgroup of order n/δ of G is made up of elements m satisfying $mn/\delta = 0$ in $\mathbf{Z}/n\mathbf{Z}$. There are exactly n/δ elements satisfying this relation, the $\alpha\delta$ with $0 \leq \alpha < n/\delta$. We therefore obtain a single group of order n/δ.

(2) This is from the course.

(3) We have $(\mathbf{Z}/p\mathbf{Z})^*$ cyclic. The result is therefore a consequence of (1) and of the Galois correspondence.

(4) The Galois group being commutative, $K_d/\mathbf{Q}$ is Galois. Then $G_d = \mathrm{Gal}(K/K_d)$ by definition of K_d.

(5) Left multiplication (resp. right) by g induces a bijection of G_d. The formulas follow from this. Then

$$p_d^2 = \frac{d}{(p-1)} \sum_{g \in G_d} g p_d = \frac{d}{(p-1)} \sum_{g \in G_d} p_d = p_d.$$

So p_d is a projector. For $g \in G_d$ and $x \in K$, we have $g(p_d(x)) = p_d(x)$ so $p_d(x) \in K^{G_d} = K_d$. Now, for $x \in K_d$, we have

$$p_d(x) = \frac{d}{p-1} \sum_{g \in G_d} x = x.$$

So $\mathrm{Im}(p_d) = K_d$.

(6) We identify G and $\mathbf{Z}/(p-1)\mathbf{Z}$ by the cyclotomic character. Then $G_d = \{k \times d\}_{0 \leq k \leq (p-1)/d}$. So

$$p_d(\xi) = \frac{d}{p-1} \sum_{0 \leq k \leq (p-1)/d} \xi^{kd} = \frac{1}{p-1} \sum_{0 \leq k \leq p-1} \xi^{kd} = \frac{\xi_d}{p-1}.$$

So $\mathbf{Q}[\xi_d] \subset \mathrm{Im}(p_d) = K_d$.

(7) We have

$$K = \mathbf{Q}[\xi] = \bigoplus_{1 \leq k \leq p-1} \mathbf{Q}\xi^k = \bigoplus_{g \in G} \mathbf{Q}g(\xi).$$

(8) We have

$$p_d(g(\xi)) = g(p_d(\xi)) = g(\xi_d)/(p-1).$$

But $\mathbf{Q}[\xi_d]/\mathbf{Q}$ is Galois because G is commutative. So $g(\xi_d) \in \mathbf{Q}[\xi_d]$ and $p_d(g(\xi)) \in \mathbf{Q}[\xi_d]$. As the $g(\xi)$, $g \in G$, generate K, the $p_d(g(\xi))$ generate $\mathrm{Im}(p_d) = K_d$. As they belong to $\mathbf{Q}[\xi_d]$, we have $K_d \subset \mathbf{Q}[\xi_d]$. We have already shown the other inclusion.

(9) For $k \in (\mathbf{Z}/p\mathbf{Z})^*$ we have $(k^{\frac{p-1}{2}})^2 = 1$ so $k^{\frac{p-1}{2}} = \pm 1$. The value is 1 for $\frac{p-1}{2}$ elements and -1 for the others. So

$$\xi_{\frac{p-1}{2}} = \frac{p-1}{2}\left(\exp\left(\frac{2I\pi}{p}\right) + \exp\left(\frac{-2I\pi}{p}\right)\right) = (p-1)\cos\left(\frac{2\pi}{p}\right).$$

Hence the result.

Solution to Exercise 12.17

(1) Let $\sigma = (\bar{1}, \bar{2}, \ldots, \bar{p})$ and $\tau = (\bar{i}, \bar{j})$. For $k \in \mathbf{Z}$, we have

$$\sigma^k \tau \sigma^{-k} = (\overline{i+k}, \overline{j+k})$$

so $(\overline{i+k}, \overline{j+k}) \in G$. In particular, for $k = -i$, we have $(\bar{0}, \overline{j-i}) \in G$. So

$$(\overline{i+k(j-i)}, \overline{i+(k+1)(j-i)}) = \sigma^{i+k(j-i)}(\bar{0}, \overline{j-i})\sigma^{-(i+k(j-i))} \in G.$$

(2) For $k = 1$ the result is clear because $\tau \in G$. Then, if $(\bar{i}, \overline{i+k(j-i)}) \in G$, we obtain that $(\bar{i}, \overline{i+(k+1)(j-i)})$ is equal to

$$(\overline{i+k(j-i)}, \overline{i+(k+1)(j-i)})(\bar{i}, \overline{i+k(j-i)})$$
$$\times (\overline{i+k(j-i)}, \overline{i+(k+1)(j-i)}) \in G.$$

(3) As $j - i \in [1, p-1]$, its image $\overline{j-i} \neq 0$ is invertible in the field $(\mathbf{Z}/p\mathbf{Z})$. It suffices to consider its inverse $\bar{k} \in (\mathbf{Z}/p\mathbf{Z})^*$.
(4) We choose $\bar{k}$ as in the previous question and conclude with 2. Then we write

$$(\bar{t}, \overline{t+1}) = \sigma^{t-i}(\bar{i}, \overline{i+1})\sigma^{i-t}.$$

(5) According to (2), we have $(\bar{t}, \overline{t+k}) \in G$ for all $k \in \mathbf{Z}$. Therefore, all transpositions are in G. As they generate S_p, we have $G = S_p$.
(6) By renumbering $\{1, \ldots, p\}$ if necessary, we can assume that $c = \sigma$. The result therefore follows from 5.
(7) In S_4, study the subgroup generated by $(1, 2, 3, 4)$ and $(1, 3)$.

Solution to Exercise 12.18

(1) As P is not irreducible, it has an irreducible factor S of degree $\leq n/2$. Then S, and therefore P, has a root in $k[X]/(S)$ with

$$[k[X]/(S) : k] = \deg(S) \leq n/2.$$

(2) Let us first note that $[\mathbf{F}_{p^d} : \mathbf{F}_p] = d$. So if P is reducible, according to (1) it has a root in a field of degree $\leq n/2$ over $\mathbf{F}_p$, that is to say in a finite field $\mathbf{F}_{p^d}$ with $d \leq n/2$. Now, if P is irreducible, let $x \in \overline{\mathbf{F}_p}$ be a root of P. Then the minimal

polynomial of x over $\mathbf{F}_p$ is P, so $\mathbf{F}_p[x]$ is of degree n over $\mathbf{F}_p$. So x is not in a field $\mathbf{F}_{p^d}$ with $d \le n/2$.

(3) We have

$$\mathbf{F}_{p^d} = \{x \in \overline{\mathbf{F}_p} \,|\, x^{p^d} - x = 0\}.$$

So saying that P has no root in $\mathbf{F}_{p^d}$ is equivalent to saying that

$$P \wedge (X^{p^d} - X) = 1.$$

Solution to Exercise 12.19

(1) We have $P'(X) = 5X^4 - 1$. If x were a common root of P and P', we would have $x/5 - x + 3 = 0$ and $x = 15/4$, which is clearly not a root of P'. So P is separable. The faithful action of G on the 5 distinct roots of P provides an embedding of G into S_5.

(2) We have in $\mathbf{F}_3[X]$,

$$\overline{P} = X^5 - X = X(X^4 - 1) = X(X - 1)(X + 1)(X^2 + 1).$$

The polynomial $X^2 + 1$ is irreducible in $\mathbf{F}_3[X]$ because it has no root in $\mathbf{F}_3$.

(3) Let $k = \mathbf{F}_5$ and $n = 5$. For $1 \le d \le n/2$, we have $d = 1$ or $d = 2$. For $d = 1$, a root $X^5 - X$ in $\overline{\mathbf{F}_5}$ cannot be a root of $X^5 - X + 3$ because $3 \ne 5$. So

$$\mathrm{GCD}(X^5 - X, X^5 - X + 3) = 1.$$

For $d = 2$, suppose we have a common root x of $X^{25} - X$ and $X^5 - X + 3$ in $\overline{\mathbf{F}_5}$. Then by applying the Frobenius morphism, we get

$$x^{25} = x^5 - 3^5 = x^5 - 3 = x - 6.$$

But then $x - 6 = x$ and $1 = 0$, a contradiction. So

$$\mathrm{GCD}(X^{25} - X, X^5 - X + 3) = 1.$$

By applying question (2), we find that $X^5 - X + 3$ is irreducible in $\mathbf{F}_5[X]$.

(4) The reduction modulo p theorem applies for $p = 3$ and $p = 5$. For $p = 3$ we get a transposition (action on the roots of $X^2 + 1$). For $p = 5$ we obtain a 5-cycle. The result then follows by question (1).

(5) As the modulo 5 reduction of P is irreducible, P is irreducible in $\mathbf{Z}[X]$. Moreover, P is monic, so Gauss's Lemma applies and P is irreducible in $\mathbf{Q}[X]$.

(6) Suppose that a root x of P is contained in $\mathbf{Q}[\exp(\frac{2I\pi}{n})]$. We then have

$$\mathbf{Q} \subset \mathbf{Q}[x] \subset \mathbf{Q}[\exp(\frac{2I\pi}{n})].$$

However, $\mathbf{Q}[\exp(\frac{2I\pi}{n})]/\mathbf{Q}$ is Galois with a commutative Galois group. Therefore, $\mathbf{Q} \subset \mathbf{Q}[x]$ is Galois with a commutative Galois group. We then have

$$\text{Gal}(\mathbf{Q}[x]/\mathbf{Q}) = \text{Gal}(P, \mathbf{Q}) = G,$$

a contradiction because S_5 is not commutative.

Solution to Exercise 12.20

(1) As g preserves distances, if g fixes two consecutive vertices it fixes all the vertices.

(2) The group Γ is a strict subgroup of $\text{Bij}(C)$ according to the previous question. According to Lagrange's theorem, its order is therefore at most $16/2 = 8$. The 8 given elements are clearly in Γ and distinct, hence the equality. We then easily verify that $\rho\sigma\rho\sigma$ fixes two consecutive elements of C.

(3) We have seen that Γ is of order 8. If Γ was abelian, we would have $\rho^{-1} = \rho\sigma^2 = \rho$, which is false. Now $\langle\rho\rangle$ forms a cyclic subgroup of order 4 of Γ. As $\sigma^{-1}\rho\sigma = \rho$, we find that it is a normal subgroup of Γ. The quotient is of order 2, so it can only be $\mathbf{Z}/2\mathbf{Z}$. We therefore obtain the exact sequence requested.

(4) We first have the two trivial subgroups $\{1\}$ and Γ, both normal. According to Lagrange's theorem, the other subgroups are of order 2 or 4. The elements of order 2 of Γ are

$$\rho^2, \sigma, \sigma\rho, \sigma\rho^2, \sigma\rho^3.$$

This gives 5 subgroups of order 2. We see that only $\langle\rho^2\rangle$ is normal among them. Then, $\langle\rho\rangle$, $\langle\sigma, \rho^2\rangle$ are the subgroups of order 4 and are normal.

(5) We have $L \subset \mathbf{R}$, but x has conjugates over $\mathbf{Q}$ that are non-real, for example ix. Galois's theorem implies that $\text{Gal}(L, \mathbf{Q}[x])$ is non-normal in G, and therefore G is non-abelian.

(6) We have

$$K = \mathbf{Q}[x, Ix, -x, -Ix] = \mathbf{Q}[x, I].$$

We have $L \subset K$ and $L \neq K$ because $L \subset \mathbf{R}$. Moreover, $K = L[I]$ and I are of degree 2 over L, so

$$[K : \mathbf{Q}] = 2[L : \mathbf{Q}].$$

As $X^4 - 2$ cancels x, we have $[L : \mathbf{Q}] = 2$ or 4. If it is 2, we have $L = \mathbf{Q}[\sqrt{2}]$. So $x = \alpha + \beta\sqrt{2}$ with $\alpha, \beta \in \mathbf{Q}$ and

$$\sqrt{2} = \alpha^2 + 2\beta^2 + 2\alpha\beta\sqrt{2}.$$

So $\alpha^2 + 2\beta^2 = 0$ and $\alpha = \beta = 0$, a contradiction. As $K/\mathbf{Q}$ is Galois, we get

$$|G| = [K : \mathbf{Q}] = 8.$$

(7) We can choose C whose vertices are $\{x, -x, \mathrm{I}x, -\mathrm{I}x\}$. Then G acts on the vertices of C. Moreover, for α, β two opposite vertices, we have $\alpha = -\beta$ and therefore $\sigma(\alpha) = -\sigma(\beta)$ for $\sigma \in G$. Thus α, β are opposite if and only if $\sigma(\alpha)$ and $\sigma(\beta)$ are. We deduce that σ preserves the distances between the vertices. We therefore have an injective morphism $G \to \Gamma$, which is an isomorphism by cardinality.
(8) We take $r = \rho$. Then $r(x) = \mathrm{I}x$ and $r(\mathrm{I}x) = -x$, which implies $r(\mathrm{I}) = \mathrm{I}$.
(9) We take $s = \sigma$ to be symmetry with respect to the diagonal $(x, -x)$. Then $s(x) = x$ and $s(\mathrm{I}x) = -\mathrm{I}x$ which implies $s(\mathrm{I}) = -\mathrm{I}$.
(10) We have already seen the relation in 2 and that s, r generate the group.
(11) We use the Galois theorem and question (4).
(12) It suffices to consider an element that is not in one of the subfields found above, that is to say not invariant under any of the elements of G. For example, $x + \mathrm{I}$ indeed has 8 conjugates

$$\{x + \mathrm{I}, \mathrm{I}x + \mathrm{I}, -x + \mathrm{I}, -\mathrm{I}x + \mathrm{I}, x - \mathrm{I}, \mathrm{I}x - \mathrm{I}, -x - \mathrm{I}, -\mathrm{I}x - \mathrm{I}\}.$$

Solution to Exercise 12.21

(1) We have $P'(X) = 5X^4 - 10X$. Therefore a common complex root x of P and P' satisfies $x^3 = 2$ and $2x^2 - 5x^2 = -1$ so $x^2 = 1/3$, a contradiction. Therefore P is separable, and from the course we obtain the morphism (faithful action of the Galois group on the roots of the polynomial).
(2) We have $\overline{P}(X) = X^5 + X^2 + 1$. So $\overline{P}(0) = 1 \neq 0$, $\overline{P}(1) = 1 \neq 0$.
(3) Let $x \in \mathbf{F}_4$ be a root of $\overline{P}$. Then $x \neq 0$ and $x^3 = 1$, so $x^2 + x^2 + 1 = 0$ and $1 = 0$, a contradiction.
(4) Suppose that $\overline{P}$ is not irreducible. Then it has an irreducible factor S of degree less than or equal to 2, so it must be equal to 2 because $\overline{P}$ does not have root in $\mathbf{F}_2$. Then $\mathbf{F}_2[X]/(S) \simeq \mathbf{F}_4$ and $\overline{P}$ has a root in $\mathbf{F}_4$, a contradiction.
(5) By reduction modulo 2, we see that the monic P is irreducible in $\mathbf{Z}[X]$. It is therefore in $\mathbf{Q}[X]$ according to the course.
(6) Let x be a root of P in the splitting field K of P. Then $\mathbf{Q}[x] \subset K$ and $[\mathbf{Q}[x] : \mathbf{Q}] = 5$ because P is the minimal polynomial of x. According to the telescopic base theorem, 5 divides $[K : k]$, which equals $|G|$ because K/k is Galois.
(7) It suffices to decompose such an element into a product of cycles with disjoint supports.
(8) For $p = 2$, the reduction modulo p theorem applies.
(9) We study the map $f(x) = P(x)$ from $\mathbf{R}$ to $\mathbf{R}$. The derivative $5X^4 - 10X$ has two real roots, 0 and $\sqrt[3]{2}$. So P has at most 3 real roots. But $P(0) = 1 > 0$ and $P(1) = -3 < 0$, so P has three real roots.

(10) As P has coefficients in **R**, the transposition is in G. It fixes the three real roots, and exchanges the two non-real conjugate complex roots. Therefore, it is a transposition.

(11) According to the previous questions, $G = S_5$ and $|G| = 5!$. As S_5 is not a solvable group, the equation is not solvable by radicals.

Solution to Exercise 12.22

(1) Course.

(2) We have $P(X) = (X^m - 1)^2$. If $a = 1$, the roots of P are the mth roots of 1 so they are in k and $K = k$.

(3) As

$$X^{2m} - 1 = (X^m - 1)(X^m + 1),$$

the mth roots of 1 and -1 are $(2m)^{th}$ roots of 1. As $(X^2 - \epsilon)' = 2X$ in characteristic $\neq 2$, the polynomial $X^2 - \epsilon$ is separable. Let $x_1 \neq x_2$ be these roots in an algebraic closure of k. For $2 \leq \alpha \leq m$, $(x_1^\alpha)^2 = \epsilon^\alpha$. As ϵ is primitive, the ϵ^α $(1 \leq \alpha \leq m)$ are distinct. Therefore, the set $\{x_2, x_1, x_1^2, \ldots, x_1^m\}$ has $m + 1$ elements. Therefore, at least one element of this set is a mth root of -1. If it is x_2, we are done. Otherwise, if $x_1^m = 1$, we have $(x_1^\alpha)^m = 1$ for all α. Therefore, $x_1^m = -1$.

(4) We have $P(X) = (X^m + 1)^2$. The roots of P are the mth roots of -1. If k contains a mth root of -1, it contains all of them, therefore $K = k$. Otherwise, $X^2 - \epsilon \in k[X]$ is irreducible according to question 2, and $K = k[X]/(X^2 - \epsilon)$.

(5) We have

$$P'(X) = 2mX^{2m-1} - 2amX^{m-1} = 2mX^{m-1}(X^m - a).$$

0 is not a root of P. If P and P' have a common root x in an algebraic closure of k, we have $x^m = a$ and $a^2 - 2a^2 + 1 = 0$, therefore $a = \pm 1$. In the case $a = \pm 1$, the polynomial P is a square, therefore not separable. Therefore, P is separable if and only if $a \neq \pm 1$.

(6) The solutions of $Y^2 - 2aY + 1 = 0$ are $a \pm \sqrt{a^2 - 1}$. Therefore, the roots of P are

$$\epsilon^k \sqrt[m]{a \pm \sqrt{a^2 - 1}} \text{ for } 0 \leq k \leq m - 1.$$

(7) The equation being solvable by radicals, the Galois group G is solvable.

(8) First note that

$$\sqrt[m]{a - \sqrt{a^2 - 1}} \sqrt[m]{a + \sqrt{a^2 - 1}} = \sqrt[m]{a^2 - (a^2 - 1)} = \sqrt[m]{1} \in k.$$

The roots of P are therefore of the form $(\epsilon^k x)^{\pm 1}$ with $0 \le k \le m - 1$. Therefore, $K = \mathbf{Q}[x]$. Now, P is a lower degree annihilating polynomial of x, therefore $[K : k] \le m$.

(9) We have $\deg(\pi_{x,k}) = [K : k] = 2m$. But $\pi_{x,k}|P$, so $P = \pi_{x,k}$ is irreducible in $k[X]$.

(10) Follows from question (6).

(11) Since ϵx is a root of P, it is a conjugate of x over k, hence the existence of g. Uniqueness comes from $K = k[x]$. For the base, we can choose $\{x^n\}_{0 \le n \le 2m-1}$ with respective eigenvalues $\{\epsilon^n\}_{0 \le n \le 2m-1}$.

(12) We have $g^m = 1$ and, as ϵ is primitive, $g^r \ne 1$ for $1 \le r \le m - 1$. So g is of order m in G and generates a cyclic subgroup of order m.

(13) By decomposing a vector of K^H on the basis of powers of x, we get that $K^H = k[x^m]$ extension of degree 2 of k. We have

$$x^m = a \pm \sqrt{a^2 - 1} \text{ with } (a + \sqrt{a^2 - 1})(a - \sqrt{a^2 - 1}) = 1.$$

As $\pi_{x^m,k}(X) = X^2 - 2aX^m + 1$, the conjugates of x^m over k are x^m and x^{-m} therefore in K^H. So K^H/k is Galois and H is normal in G.

(14) It suffices to consider the projection map $\pi : G \to G/H \simeq \mathbf{Z}/2\mathbf{Z}$. Its kernel is exactly H.

(15) The existence and uniqueness of h is treated as in question 11). Moreover $h^2(x) = x$, so $h^2 = \mathrm{Id}$. Then $[K^{H'} : k] = m$ and $(x^k + x^{-k})_{0 \le k \le m-1}$ is a basis on k. Thus

$$K^{H'} = k[x + x^{-1}].$$

(16) We have

$$(ghg^{-1})(x) = (gh)(\epsilon^{-1}x) = \epsilon^{-2}x^{-1} = (g^2 h)(x).$$

So if $\epsilon^{-2} \ne 1$, that is if $m \ge 3$, we have $ghg^{-1} \notin H'$ and H' is not normal in G. If $m = 1$, we have $G = H'$ and H' is normal in G. If $m = 2$, we have

$$(x + x^{-1})^2 = 2 + x^2 + x^{-2} = 2(a + 1) \in k.$$

So the conjugates in $x + x^{-1}$ over k are $x + x^{-1}$ and $-(x + x^{-1})$ in $K^{H'}$. The extension $K^{H'}/k$ is therefore Galois and H' is normal in G.

   In conclusion, H' is distinguished in G if and only if $m \le 2$.

(17) From the previous question, G is not commutative if $m \ge 3$. If $m = 1$, G is of order 2 so $G \simeq \mathbf{Z}/2\mathbf{Z}$. If $m = 2$, g and h commute. So $G \simeq (\mathbf{Z}/2\mathbf{Z}) \times (\mathbf{Z}/2\mathbf{Z})$ is commutative.

Solution to Exercise 12.23

(1) For $a = -1$, $n = 4$ is suitable. For $a = 2$, $n = 8$ is suitable because $\sin(\pi/4) = \sqrt{2}/2$.

(2) We have

$$\xi + \cdots + \xi^{p-1} = \xi \frac{1 - \xi^{p-1}}{1 - \xi} = \xi \frac{1 - \xi^{-1}}{1 - \xi} = -1.$$

We have $[K : \mathbf{Q}] = p - 1$ and $K = \mathbf{Q}[\xi]$. So $\{1, \ldots, \xi^{p-2}\}$, or indeed $\{\xi, \ldots, \xi^{p-1}\}$ is a $\mathbf{Q}$-basis of K.

(3) We then have

$$- z(\xi + \cdots + \xi^{p-1}) = a_1\xi + \cdots + a_{p-1}\xi^{p-1}.$$

By identifying the coefficients in $\mathbf{Q}$, we obtain the result.

(4) First, H is clearly a subgroup of G. The map $G \to H$, $g \mapsto g^2$ is a surjective group morphism whose kernel is $\{1, -1\}$. Therefore $|H| = (p - 1)/2$.

(5) As all the ξ^j ($1 \le j \le p - 1$) do not appear in the expression for $x \in \mathcal{B}$, the result follows from question (3). The Galois correspondence implies that K^H is an extension of $\mathbf{Q}$ of degree 2. We also clearly have $x \in K^H$. Hence the result.

(6) We have $|g \cdot H| = (p - 1)/2$. Moreover $g \cdot H \subset \overline{H}$. Hence the equality. Now, for $g \in \overline{H}$, we have $g(x) = \sum_{g \in \overline{H}} g(\xi)$. Hence the result.

(7) Let $x' \ne x$ be a conjugate of x. We have $x + x' = \sum_{g \in G} g(\xi) = -1$. But $P(X) = (X - x)(X - x') = \pi_{\mathbf{Q},x}(X) \in \mathbf{Q}[X]$ and $xx' \in \mathbf{Q}$. But $P(x) = 0 = x^2 - x(x + x') + xx' = x^2 + x + xx'$, so $x^2 + x \in \mathbf{Q}$.

(8) We write

$$x^2 + x = \sum_{g \in H} g(\xi) + \sum_{g, g' \in H} g(\xi)g'(\xi).$$

Each $g(\xi) \in \mathcal{B}$. Each $g(\xi)g'(\xi)$ is in $\mathcal{B}$, or equals 1. But $g(\xi)g'(\xi) = 1$ is equivalent to $g = -g'$ in $\mathcal{F}_p$. As $-1 \in H$, this occurs $(p - 1)/2$ times. We therefore obtain the result with

$$\sum_i a_i = \frac{p - 1}{2} + \frac{(p - 1)^2}{4} - \frac{p - 1}{2} = \frac{(p - 1)^2}{4}.$$

By applying (3), we get

$$x^2 + x - \frac{p - 1}{2} = -\frac{(p - 1)^2}{4(p - 1)} = -\frac{p - 1}{4}$$

and $x^2 + x = (p - 1)/4$. By solving the equation, we get $x = (-1 \pm \sqrt{p})/2$ and the result.

(9) We work in the same way. As $-1 \notin H$, $g(\xi)g'(\xi) = 1$ does not occur. So we get the result with

$$\sum_i a_i = \frac{p-1}{2} + \frac{(p-1)^2}{4} = \frac{(p-1)(p+1)}{4}.$$

So

$$x^2 + x = -\frac{(p-1)(p+1)}{4(p-1)} = -\frac{p+1}{4}.$$

We get $x = (-1 \pm \sqrt{-p})/2$ and the result.

(10) We have shown the result for $\sqrt{p}$ or $\sqrt{-p}$. We deduce the result for $\sqrt{n}$ where n or $-n$ is a power of a prime number. We conclude in general by noting that for all p, q, we have $\mathbf{Q}(e^{2l\pi/p}, e^{2l\pi/q}) \subset \mathbf{Q}(e^{2l\pi/pq})$.

References

[Bou89] N. Bourbaki, *Elements of Mathematics, Algebra I*, Chap. 1–3, Springer, 1989

[Bou23] N. Bourbaki, *Elements of Mathematics, Algebra II*, Chap. 4-7, Springer, 2003

[Cha05] A. Chambert-Loir, *A Field Guide to Algebra*, Springer, 2005

[Dou20] R. and A. Douady, *Algebra and Galois Theories*, Springer, Cham, 2020

[Ehr11] C. Ehrhardt, Le bicentenaire d'Évariste Galois (1811–1832), *Gaz. Math.* No. **129** (2011), 71–73

[Elk02] R. Elkik, *Cours d'Algèbre*, Ellipses Marketing Col.: University Mathematics, 2002

[Gal62] E. Galois, *Écrits et mémoires mathématiques d'Évariste Galois*, Gauthiers-Villars, 1962

[Gro71] A. Grothendieck, *Séminaire de Géométrie Algébrique du Bois Marie – 1960–61 – Revêtements étales et groupe fondamental (SGA 1)*, Springer-Verlag, 1971

[Lan02] S. Lang, *Algebra*, Graduate Texts in Mathematics **211**, Springer-Verlag, 2002

[Mun00] J. Munkres, *Topology*, Prentice Hall, Inc., Upper Saddle River, NJ, 2000

[Rud87] W. Rudin, *Real and Complex Analysis*, McGraw-Hill Book Co., New York, 1987

[Ser87] J.-P. Serre, Groupe de Galois sur **Q**, *Bourbaki Seminar*, **30** (1987–1988), Lecture No. 689

[Ste15] I. Stewart, *Galois theory*, CRC Press, Boca Raton, FL, 2015

[Vol92] H. Völklein, $GL_n(q)$ as Galois group over the rationals, *Mathematical Annals* (1992), vol. **293**, no. 1, 163–176

[Wae49] B.L.H. van der Waerden, *Modern Algebra. Vol. I*, Frederick Ungar Publishing Co., New York, N. Y., 1949

© The Author(s), under exclusive license to Springer Nature Switzerland AG 2024
D. Hernandez, Y. Laszlo, *Introduction to Galois Theory*, Springer Undergraduate
Mathematics Series, https://doi.org/10.1007/978-3-031-66182-2

Index

© The Author(s), under exclusive license to Springer Nature Switzerland AG 2024
D. Hernandez, Y. Laszlo, *Introduction to Galois Theory*, Springer Undergraduate
Mathematics Series, https://doi.org/10.1007/978-3-031-66182-2